World Geomorphological Landscapes

Series editor

Piotr Migoń, Wrocław, Poland

For further volumes:
http://www.springer.com/series/10852

Stefan Grab • Jasper Knight
Editors

Landscapes and Landforms of South Africa

Springer

Editors
Stefan Grab
School of Geography, Archaeology
and Environmental Studies
University of the Witwatersrand
Johannesburg
South Africa

Jasper Knight
School of Geography, Archaeology
and Environmental Studies
University of the Witwatersrand
Johannesburg
South Africa

ISSN 2213-2090 ISSN 2213-2104 (electronic)
World Geomorphological Landscapes
ISBN 978-3-319-03559-8 ISBN 978-3-319-03560-4 (eBook)
DOI 10.1007/978-3-319-03560-4

Library of Congress Control Number: 2015931450

Springer Cham Heidelberg New York Dordrecht London

Printed on acid-free paper

Springer International Publishing AG Switzerland is part of Springer Science+Business Media
(www.springer.com)

Series Editor Preface

Landforms and landscapes vary enormously across the Earth, from high mountains to endless plains. At a smaller scale, nature often surprises us creating shapes which look improbable. Many physical landscapes are so immensely beautiful that they received the highest possible recognition—they hold the status of World Heritage Sites. Apart from often being immensely scenic, landscapes tell stories which not uncommonly can be traced back in time for tens of million years and include unique geological events such as meteorite impacts. In addition, many landscapes owe their appearance and harmony not solely to the natural forces. For centuries, and even millennia, they have been shaped by humans who have modified hill-slopes, river courses, and coastlines, and erected structures which often blend with the natural landforms to form inseparable entities.

These landscapes are studied by geomorphology—'the science of scenery'—a part of Earth Sciences that focuses on landforms, their assemblages, surface, and subsurface processes that molded them in the past and that change them today. To show the importance of geomorphology in understanding the landscape, and to present the beauty and diversity of the geomorphological sceneries across the world, we have launched a book series *World Geomorphological Landscapes*. It aims to be a scientific library of monographs that present and explain physical landscapes, focusing on both representative and uniquely spectacular examples. Each book will contain details on geomorphology of a particular country or a geographically coherent region. This volume presents the geomorphology of South Africa—a country that not only hosts superb and highly diverse landforms and landscapes—basaltic plateaus, imposing escarpments, inselbergs, intriguing sandstone formations, waterfalls, pans, and dunes, but can also be considered as one of the inspirations for modern geomorphological studies, especially that focus on long-term landform evolution. Landscape evolution models associated with workers such as Lester King, which profoundly influenced the thinking of many mid-twentieth century geomorphologists, have been developed in South Africa. In more recent times, since the 1990s, it was South Africa where their reappraisal has been attempted through cosmogenic dating.

The World Geomorphological Landscapes series is produced under the scientific patronage of the International Association of Geomorphologists (IAG)—a society that brings together geomorphologists from around the world. The IAG was established in 1989 and is an independent scientific association affiliated with the International Geographical Union (IGU) and the International Union of Geological Sciences (IUGS). Among its main aims are to promote geomorphology and to foster dissemination of geomorphological knowledge. I believe that this lavishly illustrated series, which keeps to the scientific rigor, is the most appropriate means to fulfill these aims and to serve the geoscientific community. To this end, my great thanks go to Profs. Stefan Grab and Jasper Knight for adding this book to their agenda, successfully coordinating the large team of authors, and delivering such an exciting illustrated story to read and admire. I hope they are as pleased with the final outcome as I am. I also acknowledge the excellent work of all individual authors who accepted to share their expert knowledge of the country with the global geomorphological community. I once happened to spend a day at the foot of the Drakensberg Escarpment, which was an unforgettable experience. Now I have nearly 20 other places in South Africa to visit. I am sure readers of this volume will be equally tempted to see these marvels for themselves.

Piotr Migoń

Preface

The landforms and landscapes of South Africa present a long and diverse geological history and have been of interest to geologists and geomorphologists since Charles Darwin's visit to the Western Cape region in the 1830s. Later, the discovery and then exploitation of many metaliferous and mineral deposits including gold, platinum, and diamonds brought geologists and engineers who, as a result of mining activities, have transformed landscapes in many parts of the country but also inspired much scientific investigation, creating a geomorphological legacy that is still present today. Much work on the landforms and landscapes of South Africa has thus been achieved by incomers from Europe; here, we intend to showcase South African geomorphology to the world and provide the context for inspiring, empowering, and training a new generation of South African students and scientists to make such studies their own—an aim we try to set forth in our own research and teaching at the University of the Witwatersrand, Johannesburg.

South Africa's landscapes are very diverse, not only in their geological history, but in terms of geomorphological processes, ecosystems, land use, and relationships to cultural patterns. These relationships are explored in many chapters in this book, and are explicit in the inscription of many South African landscapes as UNESCO World Heritage Sites, including Mapungubwe Cultural Landscape (inscribed 2003), Richtersveld Cultural and Botanical Landscape (2007), and Maloti-Drakensberg Park (2000). We describe the relationships between different landscape components, and the context of different chapters in this book, in our introductory Chap. 1, to which readers are referred.

We thank all of the chapter authors and peer-reviewers of these chapters for their hard work and excellent contributions. We also thank series editor Piotr Migoń for his encouragement and rigor, Wendy Phillips for drawing many maps, and the National Research Foundation (South Africa) for supporting our South African geomorphological research.

Stefan Grab
Jasper Knight

Contents

Short Biodata of Authors

Dr. Greg Botha is a Senior Specialist Scientist at the Council for Geoscience, specializing in the Cenozoic evolution of southern Africa. His research into the lithostratigraphy and geochronology of the aeolian dunes and wetlands of Maputaland has revealed the influence of sea level fluctuations and global climate change on the evolution of the coastal plain. As a consultant he contributed to the environmental management frameworks for the region and plans for the future management of the St. Lucia estuary.

R. Grant Cawthorn is the Platinum Industry's Professor of Igneous Petrology in the School of Geosciences, University of the Witwatersrand; recipient of the first University Convocation Distinguished Teacher's Award and two Jubilee medals from the Geological Society of South Africa, and has over 140 scientific publications, mainly on the Bushveld Complex.

John S. Compton is Associate Professor in the Department of Geological Sciences at the University of Cape Town. His research focus is on the marine geology of the western and southern margins of South Africa, looking at the sediment records for clues on how these margins have evolved since the Neogene.

Dr. Frank D. Eckardt is a geomorphologist at the University of Cape Town, with 20 years of southern African field experience, including Botswana and Namibia. His interests include southern African landscape evolution, as well as aeolian and evaporitic process geomorphology. He holds a DPhil from Oxford, an MSc in Applied Remote Sensing from Cranfield, and a BSc from Kings College London.

Amanda B. Esterhuysen is Associate Professor of Archaeology in the School of Geography, Archaeology and Environmental Studies, University of the Witwatersrand. Her research interests include contemporary archaeology and the management of heritage sites. She has published a number of scientific articles and cultural resource management reports.

Roger L. Gibson is Professor of Structural Geology and Metamorphic Petrology and Head of the School of Geosciences at the University of the Witwatersrand. His research focuses on structural/tectonic and metamorphic processes in mountain belts and ancient exhumed crust, and impact structures. He co-authored, with Uwe Reimold, the successful book *Meteorite Impact!* and two scientific guide books on the geology of the Vredefort impact structure, and has published over 90 peer-reviewed scientific articles. Outside of his research, he is involved in several initiatives to develop the geotourism potential of the Vredefort Dome.

Stefan Grab is Professor of Physical Geography in the School of Geography, Archaeology and Environmental Studies, University of the Witwatersrand. His primary research interests include cold region and sandstone geomorphology, mountain environments, and climate change. He has published over 80 peer-reviewed scientific articles.

Dr. Stephan Harrison is a climate scientist working in the College of Life and Environmental Science at Exeter University in England. His research is on the response of glaciated mountain systems to climate change, and he has published over 100 peer-reviewed scientific papers.

Peter Holmes is a retired professor, and currently a research associate in the Department of Geography at the University of the Free State, Bloemfontein. He is co-editor of the publication *Southern African Geomorphology; Recent Trends and New Directions* (2012). His research interest is in aeolian sediments and associated landforms in semi-arid and coastal environments.

Jasper Knight is Professor of Physical Geography in the School of Geography, Archaeology and Environmental Studies, University of the Witwatersrand, South Africa. His research focuses on the dynamics of glacial, mountain, and coastal environments in response to climate change. He has published over 80 peer-reviewed scientific articles based on research in diverse locations globally, and many books.

Terence S. McCarthy is Professor Emeritus in the School of Geosciences. He has wide research interests and has published extensively in the areas of geology, geochemistry, geomorphology, and environmental science, especially wetlands.

Michael E. Meadows is Professor and Head of the Department of Environmental and Geographical Science at the University of Cape Town. His research interests include Quaternary environmental change and the historical impact of humans on the environment of semi-arid southern Africa. Although hardly a connoisseur, he admits to being rather fond of a glass or two of the Cape's finest sauvignon blanc as the sun goes down.

David J. Nash is Professor of Physical Geography at the University of Brighton and Honorary Research Fellow at the University of the Witwatersrand. His research mainly concerns the geomorphology and historical climatology of southern Africa, with the majority of his work focusing on the Kalahari environment. He has published over 85 journal papers and book chapters and has edited research volumes including *Geochemical Sediments and Landscapes* (Wiley 2007) and *Quaternary Environmental Change in the Tropics* (Wiley 2013).

Dr. David G. Passmore is a Lecturer in Geography in the Department of Geography, University of Toronto Mississauga, Canada. He is a geoarchaeologist with a current focus on military geography and conflict archaeology, and has published over 60 peer-reviewed scientific articles and texts.

Dr. Lynne J. Quick recently completed her Ph.D. in paleoenvironmental science at the University of Cape Town. Her research focuses on the development and evaluation of new paleoenvironmental records in southern Africa, with particular reference to the southwestern Cape and the Fynbos Biome. Her MSc thesis was centered on the Late Quaternary palaeoenvironments and vegetation change in the Cederberg and has therefore spent a lot of time working (and playing) in the Cederberg region.

David L. Reid is Professor of Geology in the Department of Geological Sciences, University of Cape Town, South Africa. He has published widely in the fields of Igneous Petrology, Economic Geology, and Precambrian Geology. A particular interest is in the geological evolution of the Proterozoic mobile belts in the SW parts of Africa, namely the Namaqua and Pan-African provinces.

W. Uwe Reimold spent 20 years at the University of the Witwatersrand before taking up the Chair of Mineralogy and Petrography at the Museum für Naturkunde Berlin and the Humboldt-Universität Berlin (Germany) in 2006. He has authored or co-authored a number of books and over 300 peer-reviewed scientific papers, many dealing with impact structures from all parts of the world. Besides impact research, he has also published widely on ore deposits and retains a strong interest in geological heritage issues.

Dr. Dominic Stratford is a research coordinator at the Sterkfontein Caves paleoanthropological site and lecturer in Geoarchaeology at The University of the Witwatersrand. His research interests are wide, spanning hominin morphology, geomorphology, sedimentology, microstratigraphy, taphonomy, and lithic technology. His main research focus is on site formation processes and the development of a broad-spectrum multidisciplinary approach toward increasing stratigraphic resolution in archeological and paleoanthropological contexts.

Dr. Stephen Tooth graduated from the University of Southampton, UK and completed a Ph.D. at the University of Wollongong, Australia. He undertook postdoctoral work at the University of the Witwatersrand, South Africa, before joining Aberystwyth University, UK. His research focuses on geomorphology, sedimentology, and environmental change, especially in the southern African and Australian drylands.

Morris Viljoen is an Emeritus Professor of Mining Geology in the Department of Geology, University of the Witwatersrand, South Africa. His research interests are in economic and environmental geology, and geoheritage promotion for education and tourism.

Landscapes and Landforms of South Africa—An Overview

1

Stefan Grab and Jasper Knight

Abstract

This chapter introduces the broader context for the diverse South African landscapes and landforms discussed in this book. We briefly summarize South Africa's long geological history, demonstrating that some of the earliest known geomorphic events on Earth are preserved as geological artefacts within the contemporary landscape. Both earlier and more recent theories of landscape evolution are then highlighted, in particular given that some of these were locally founded but applied globally. It is demonstrated that much of the South African macroscale geomorphology and site-specific landform development is controlled not only by geology, but also by past epeirogenic uplift, influencing river divides and drainage networks, and also a regionally diverse climate. Finally, the association between people and landscapes is emphasized as an important theme covered throughout this book.

Keywords

South Africa • Geology • Climate • Geomorphic evolution • Geoheritage

1.1 Introduction

South Africa boasts an extraordinarily diverse range of landscapes and landforms which are the product of an exceptionally long geologic and climatic history. While some of South Africa's landscapes are among the oldest in the world, being underlain by a Pangaea-age continental craton, other landscapes reflect contemporary geomorphological processes and the effects of ongoing climate change. In addition, the varied climate zones and ecosystems across South Africa have given rise to a very diverse contemporary land surface which reflects both past and recent geologic processes as well as the imprint of human activity over long timescales. South Africa's rich geological and mineral resources, such as gold, platinum and diamonds, have also been a primary trigger for geological investigation. Early European explorers, building from indigenous knowledge and prehistoric mineral explorations, were concerned with mapping these resources; indeed, the city of Johannesburg was founded on the wealth of gold mining of the Witwatersrand reefs (Rosenthal 1970). Other commentators have been concerned with landscape features of South Africa. For example, Charles Darwin visited Cape Town in June 1836 aboard *H.M.S. Beagle* and noted in his diary of 2 June, with respect to Table Mountain, that 'I should think so high a mountain… must be a rare phenomenon; it certainly gives the landscape a very peculiar, and from some points of view, a grand character'.

From the early twentieth century, many world-renowned geologists and geomorphologists, like Alexander du Toit, John Wellington, Lester King, Charles Twidale and Andrew Goudie, have used South Africa as a natural research laboratory for developing and testing theories of long-term landscape evolution. This includes du Toit's work on the break-up of Gondwanaland, and King's model for cycles of macroscale land surface erosion (Fig. 1.1; e.g. King 1955). This theme has continued to the present day, with examples such as models of pan evolution (Goudie and Wells 1995) or models of land surface denudation that nod to decades-old ideas of planation surfaces, but which are based here on the use of cosmogenic

S. Grab (✉) · J. Knight
University of the Witwatersrand, Johannesburg, South Africa
e-mail: stefan.grab@wits.ac.za

S. Grab and J. Knight (eds.), *Landscapes and Landforms of South Africa*,
World Geomorphological Landscapes, DOI 10.1007/978-3-319-03560-4_1,

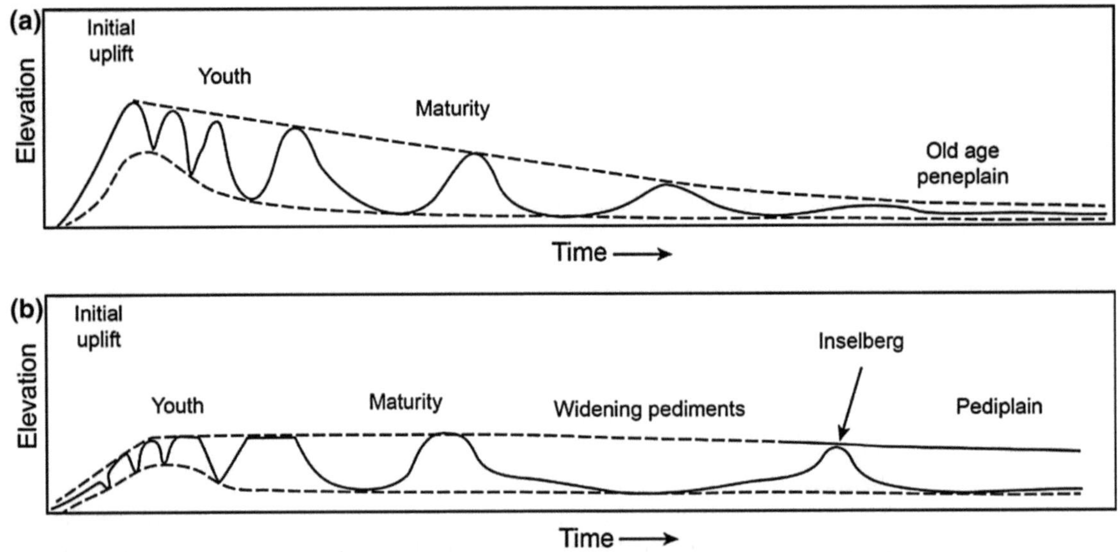

Fig. 1.1 Sketch of common landscape evolution models that have been applied to the South African landscape. **a** The Davisian model, whereby the land surface erodes downwards over time, producing a low-relief peneplain; **b** the King model whereby valleys widen over time as a result of parallel slope retreat, leading to the formation of isolated inselbergs separated by a sloping rocky pediplain (see also Fig. 1.5)

radionuclides (Bierman and Caffee 2001). In that context, therefore, studies of the geology and geomorphology of South Africa draw from a long tradition, but are increasingly focused on twenty-first-century issues of landscape sensitivity, resource management and sustainability, and the impacts of climate change (e.g. Beckedahl et al. 2002). This book helps contextualize these two complementary viewpoints of the South African landscape and its evolution.

Several previous books have included chapters outlining the geological and geomorphological evolution of the South African landscape, reflecting our understanding at the time of publication (e.g. Moon and Dardis 1988; Partridge and Maud 2000a; Holmes and Meadows 2012). This chapter will not repeat such discussions because each chapter in this book takes a geographically specific area and describes or accounts for landscape-scale geomorphology and evolution within the respective area, and does so with reference to the bigger picture of geoheritage and cultural heritage, conservation and tourism. The geographical spread of areas discussed within chapters of this book (Fig. 1.2) means that most areas of South Africa and its geodiversity are covered, including important or iconic tourist sites such as Table Mountain and Kruger National Park.

1.2 Geological Overview

Early Archean rocks exposed in the Barberton area of South Africa represent one of the world's best-known greenstone belts (Fig. 1.3). Despite their age ($\sim$2.9–3.2 Ga), these metamorphosed volcanic–sedimentary sequences, which were initially a product of oceanic spreading centres, reveal some of the earliest known geomorphic events on Earth. For instance, the Earth's oldest known siliciclastic (carbonate) tidal deposits have been described from this region (Noffke et al. 2006), as is also the oldest known glaciation (2.9 Ga; Mozaan Group), evidenced by diamictites with striated and faceted clasts contained within stratified siltstone and interpreted as ice-rafted debris (Young et al. 1998). Sedimentary rocks of the Early Proterozoic ($\sim$2.2 Ga; Pretoria Group, Magaliesberg Formation) also provide good evidence of subtidal channel systems and ephemeral braid-delta processes, which indicate high-energy macrotidal dynamics and high tidal range during deposition (Button and Vos 1977; Eriksson et al. 1995).

Following evolution of the Kaapvaal Craton, a long accretionary geological history associated with crustal extension and compression continued through to the Neogene (Partridge and Maud 2000b). This period broadly encompassed subsidence from the vertical motion of rigid basement blocks with periods of basin infills which were temporally separated by volcanic episodes. Most sedimentary rocks in South Africa are products of infilling, such as of the Main Karoo Basin and its subsidiary basins such as the Springbok Flats Basin, Ellisras Basin, Waterberg Basin and Tuli Basin. The Karoo Basin developed in south-western Gondwana during the Late Carboniferous, extending well beyond the contemporary southern African continental margin. Subsequent infilling of the basin yielded a total stratigraphic thickness of $\sim$12 km, and although much of this has been subsequently eroded, the Supergroup currently crops out over more than half of South Africa ($\sim$600,000 km^2), from the Cape fold belt (a product of active margin development during the Early Paleozoic) in the south-west to the Kaapvaal Craton

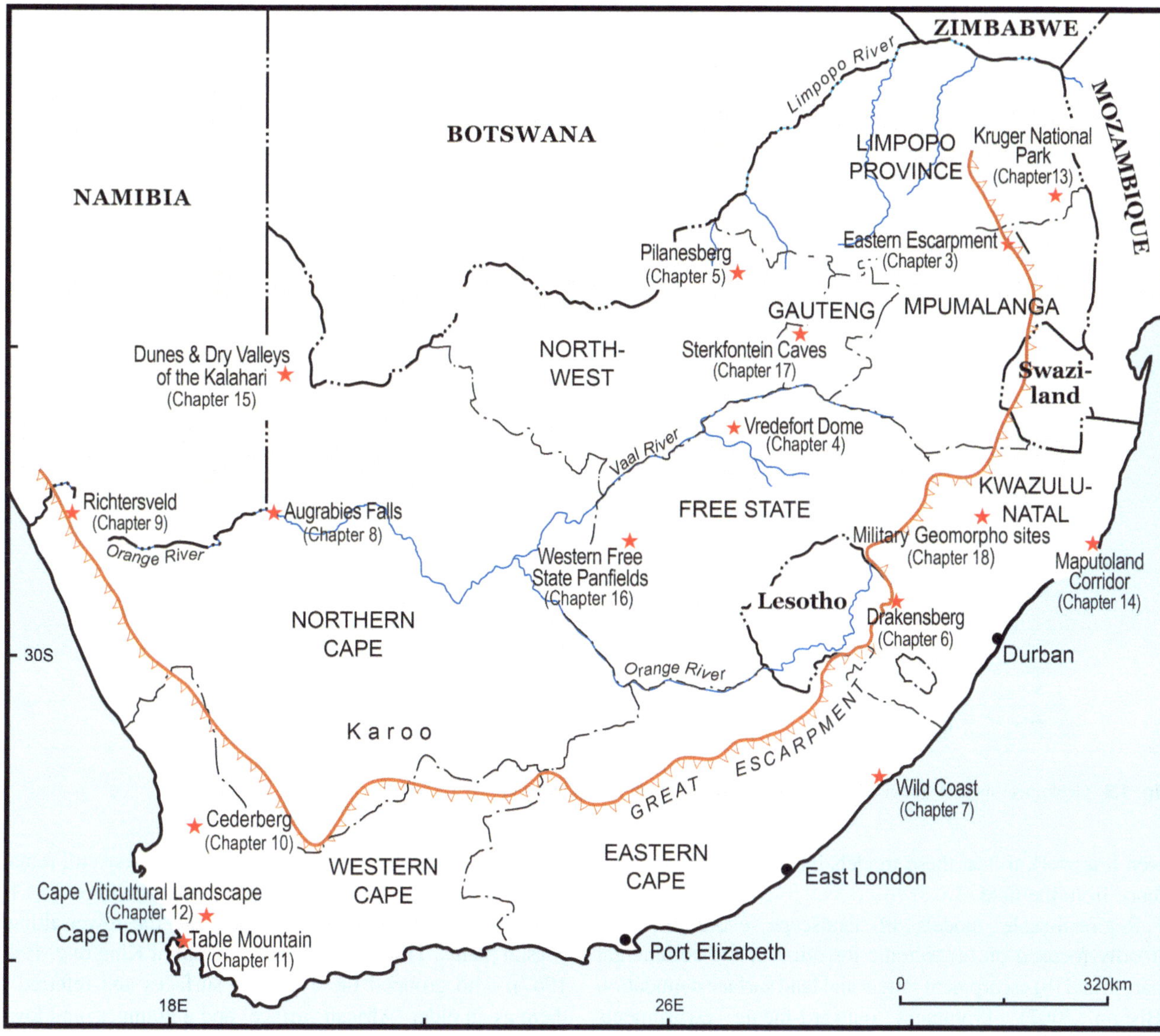

Fig. 1.2 Geographic locations of sites discussed in different chapters of this book. Numbers refer to chapter numbers (Chaps. 2 and 19 are applicable countrywide and thus not indicated here)

in the north (Smith 1995). The sandstone succession of the Karoo Basin accumulated from the Late Carboniferous ($\sim$300 Ma) to the Early Jurassic ($\sim$190 Ma) and was influenced by significant global-scale climate changes. The early Karoo foreland basin was bounded by the Cape fold belt in the south and the Cargonia Highlands in the north, and between these mountains, Dwyka-age glaciers eroded deep valleys and deposited large thicknesses of glacigenic diamicton and glaciofluvial sediments (Isbell et al. 2008). Karoo sedimentation eventually terminated during the early phases of basaltic outpourings around 183 Ma ago. Volcanicity was associated with continental rifting, which also yielded abundant dolerite dyke and sill intrusions through the entire Karoo Sequence. Volcanic activity eventually ended at $\sim$105 Ma (Eales et al. 1984). The coastline of southern Africa, similar to that known today, emerged as South America; the Falkland Plateau and Antarctica separated from southern Africa in the period $\sim$129–121 Ma (Fouché et al. 1992).

1.3 Geomorphic Evolution of the Southern African Landscape

The geologic history of South Africa and the development of its topography and landscapes are very closely linked, particularly over the Cenozoic (last 66 Ma). Debate on the topographic evolution of South Africa over this timeframe has been focused around the development of different models of landscape evolution, based on its tectonic position as a high-elevation passive continental margin. Unfortunately, there has

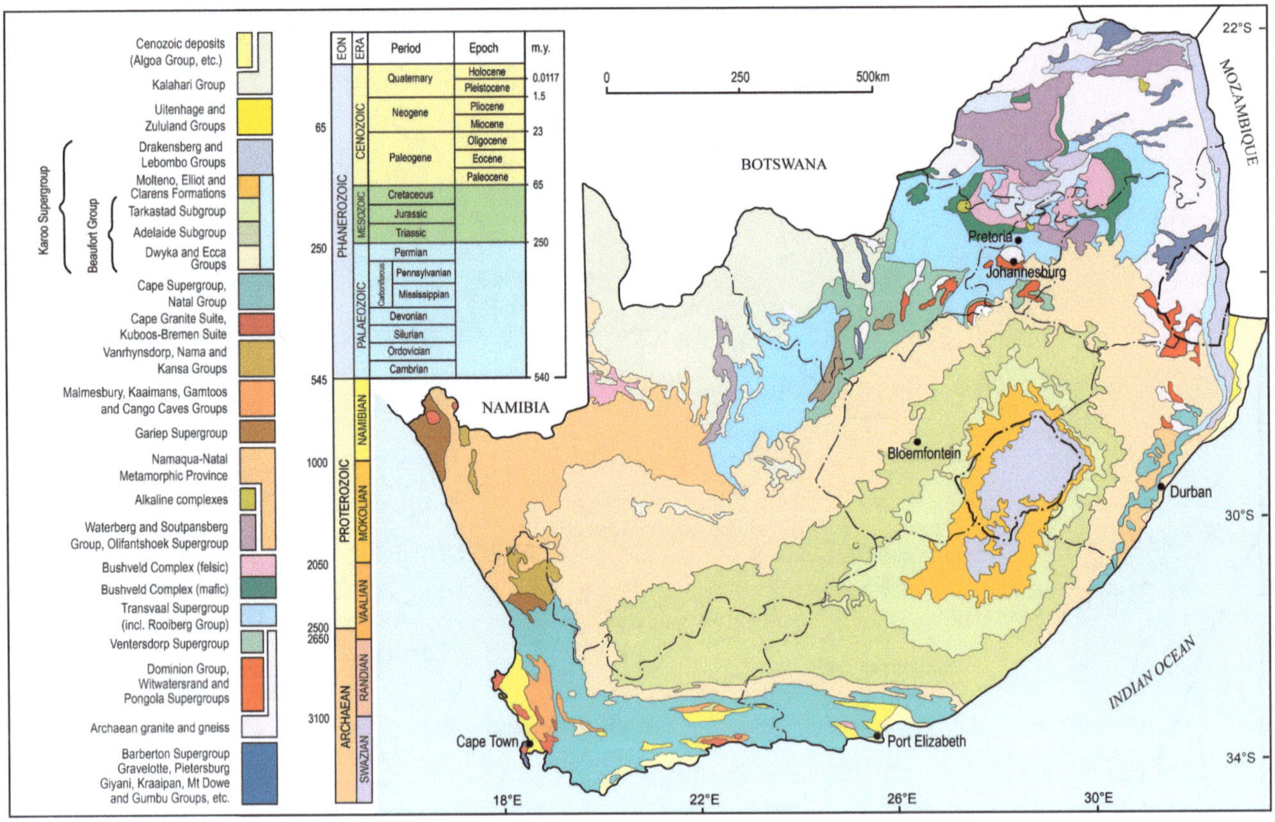

Fig. 1.3 Geological map of South Africa (modified after Council for Geoscience 1997)

been less work to test these models based on empirical evidence from the field.

Regional-scale models of landscape evolution have broadly focused on (a) tectonic lowering of the continental margin and (b) escarpment retreat and land surface denudation (Bishop 2007). Prominent outward-facing escarpments (including the Great Escarpment) that are still parallel to the southern African coastline (and are now located some 50–200 km inland; see Chaps. 3 and 6) were initially considered to be a product of lithospheric rebound following continental rifting and extensive post-Gondwana erosion of the land surface (e.g. Partridge 1998; Gilchrist and Summerfield 1990). According to Partridge (1998), humid climates during the Early Cretaceous enhanced continental weathering, erosion and development of a dense terrestrial drainage network. River erosion of the land surface helped drive backwasting of the Great Escarpment towards the continental interior and resulted in the formation of undulating or gently sloping benches across the widening coastal hinterland. Contrasting base levels of erosion were established on either side of the Great Escarpment. Landwards of this major topographic barrier, base level is provided by major rivers such as the Limpopo and Orange (or Gariep) (see Chap. 8). Seawards of the Great Escarpment, base level is that of sea level. Consequently, erosional surfaces of similar age, yet of contrasting altitude, were able to develop, and to a large extent, these still remain today and include the high plateau (pediplain; Figs. 1.1, 1.4 and 1.5) of central southern Africa and the lower-altitude coastal plains. This view differs from that of King (e.g. 1944, 1967), who grouped these erosion surfaces and referred to them as an older 'African surface' and a younger (and lower elevation) 'post-African surface'. He compared sediment build-up in coastal regions to argue that the 'African surface' graded to sea level during the Late Cretaceous to Early Miocene. King (1955) additionally proposed that such crustal erosion prompted isostatic uplift of the continental margin, thus adding elevation to the Great Escarpment. These debates on the nature of land surface evolution over large spatial and long temporal scales have been at the heart of developments in geomorphology during the twentieth century. The model of W.M. Davis involves land surface denudation by weathering and transport of weathered products along river valleys that widen over time. The resulting low-relief peneplain reflects land surface quasi-stability (Fig. 1.1a). The model of Lester King involves parallel slope retreat whereby valleys widen over time, leading to the formation of isolated inselbergs separated by a sloping rocky pediplain (Fig. 1.1b). These models, among others, provide hypotheses for land surface evolution that can be tested against field geomorphic, sedimentary and dating evidence.

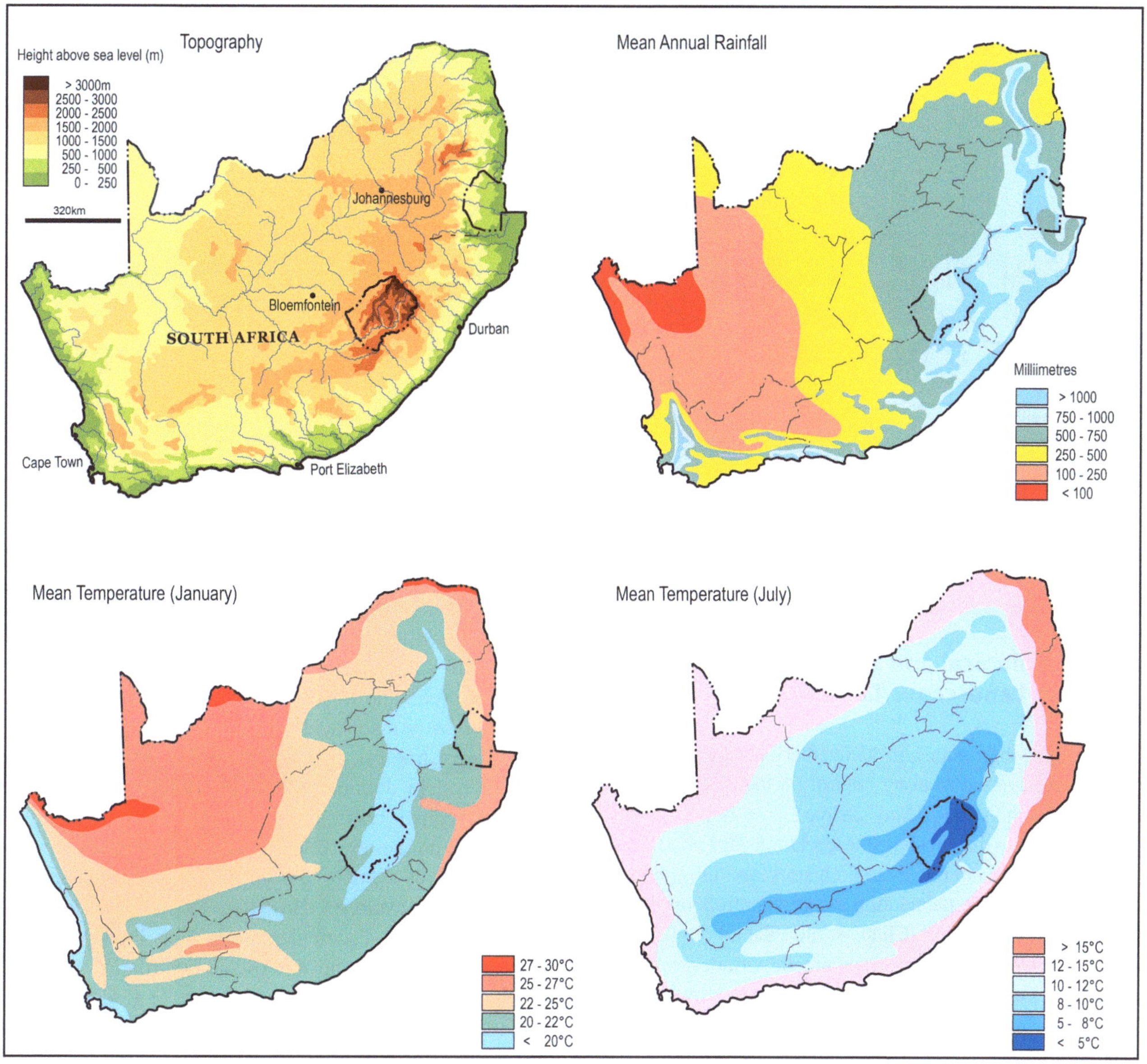

Fig. 1.4 The topography and climate of South Africa

More recently, apatite fission track (e.g. Brown et al. 2002; Tinker et al. 2008) and apatite/titanite U–Th/He geochronometric dating (Flowers and Schoene 2010) have strongly challenged the km-scale initial post-Gondwana denudation rates as argued by King, but rather propose negligible erosion since the Cretaceous. Japsen et al. (2012) use evidence of Cenozoic marine strata and the presence of tilted and truncated Upper Cretaceous sediments off the west coast of South Africa to argue that much of the contemporary land surface elevation is due to post-rift uplift. To this end, southern African drainage systems have been noted for their major reorganization since Gondwana, based on isopachs (stratigraphic thicknesses) in the Kalahari and the numerous palaeo-drainage features over the central interior (Moore and Larkin 2001). Given that major river divides cut across geological boundaries, lithology is not the only drainage control. Rather, much of the regional macroscale topography has been influenced by concentric (i.e. broadly parallel to the southern Africa coastline) epeirogenic uplift in connection with plate-boundary reorganization (Moore et al. 2009). Major uplift events during the Cenozoic have thus shifted river divides and their associated fluvial drainage networks, and consequently controlled cyclic episodes of denudation. The contemporary macroscale topography may thus be more a reflection of plate processes and dynamics (Moore et al. 2009) rather than the longer-standing idea of rising mantle plumes (e.g. Partridge 1997; Burke et al. 2008). Further radiometric dating of the land surface and exposed rocks across southern Africa will help evaluate these competing models.

Fig. 1.5 Karoo pediplain on the high plateau, with a dolerite-capped inselberg in the centre (Mountain Zebra National Park); see also Fig. 1.1 for the Karoo region (*Photograph* S. Grab)

1.4 Geological and Structural Controls on Landscapes/Landforms

The foregoing discussion considered that geology and geologic history have exerted a strong control on the macroscale evolution of South Africa. Many chapters in this book provide good case studies that illustrate these relationships on a meso- to microscale. For example, tectonic history and long-term patterns of land surface denudation are key controls on the geomorphology of Vredefort (Chap. 4) and Pilanesberg (Chap. 5). Microscale weathering processes are important geomorphic controls on sandstones (Chap. 2) and in coastal environments (Chap. 7). The chapters in this book also illustrate important geologic principles of scale (both spatial and temporal), process–form relationships, inheritance, time lags and equifinality. For example, the preservation of old plateau surfaces behind very active slopes, and the inland migration of river knickpoints (Chaps. 3 and 8) illustrate the range of rates at which weathering and erosion can take place, even within the same area. The macroscale geomorphic evolution models of King and others thus still resonate in today's South African landscapes, where geomorphic principles can be seen in action.

1.5 Climate Controls on Landscapes/Landforms

South Africa has a varied climate, both spatially and temporally (Fig. 1.4). Most of South Africa (apart from the south-western Cape) is situated within the summer rainfall zone with a distinct rainy season from November to March, and dry season from May to September. Summers are typically warm and humid along the east coast, and dry and very hot towards the western and northern regions. Summer rains are mostly associated with tropical–temperate troughs and easterly tropical air flow over the interior. A high-pressure system dominates over the interior during winter (June to August), resulting in cold, dry conditions. In contrast, the south-western Cape has a Mediterranean-type climate with wet winters associated with the passage of southerly polar cyclones. However, topographic influences such as the Great Escarpment and Cape Fold Mountains have strong regional and microscale controls on precipitation, humidity, airflow and temperature (e.g. Chaps. 9, 11 and 12). Generally, highest annual precipitation (>1,000 mm) occurs along the eastern sectors of the Great Escarpment and along the southern and eastern coastal belts, while rapid drying occurs landwards of the Great Escarpment which acts as a rain shadow. Northern and western parts of the country receive below 500 mm of precipitation annually, and most of the interior plateau has a net moisture deficit. Snow occasionally occurs along the higher mountain ranges during colder months, and severe frost (below −10 °C) is possible over the plateau in winter. South Africa thus has a diverse climatic influence on contemporary landform development, ranging from hyper-arid, hot aeolian (Chaps. 9 and 15) to cool periglacial (Chap. 6). Region-specific weathering and erosion are also strongly contrasting and climate linked (e.g. Chaps. 2, 8, 10 and 13).

South Africa has also undergone substantial regional climate fluctuations and changes, and sea level changes, through the Cenozoic (see Partridge and Maud 2000b), which geomorphologically is manifested through the

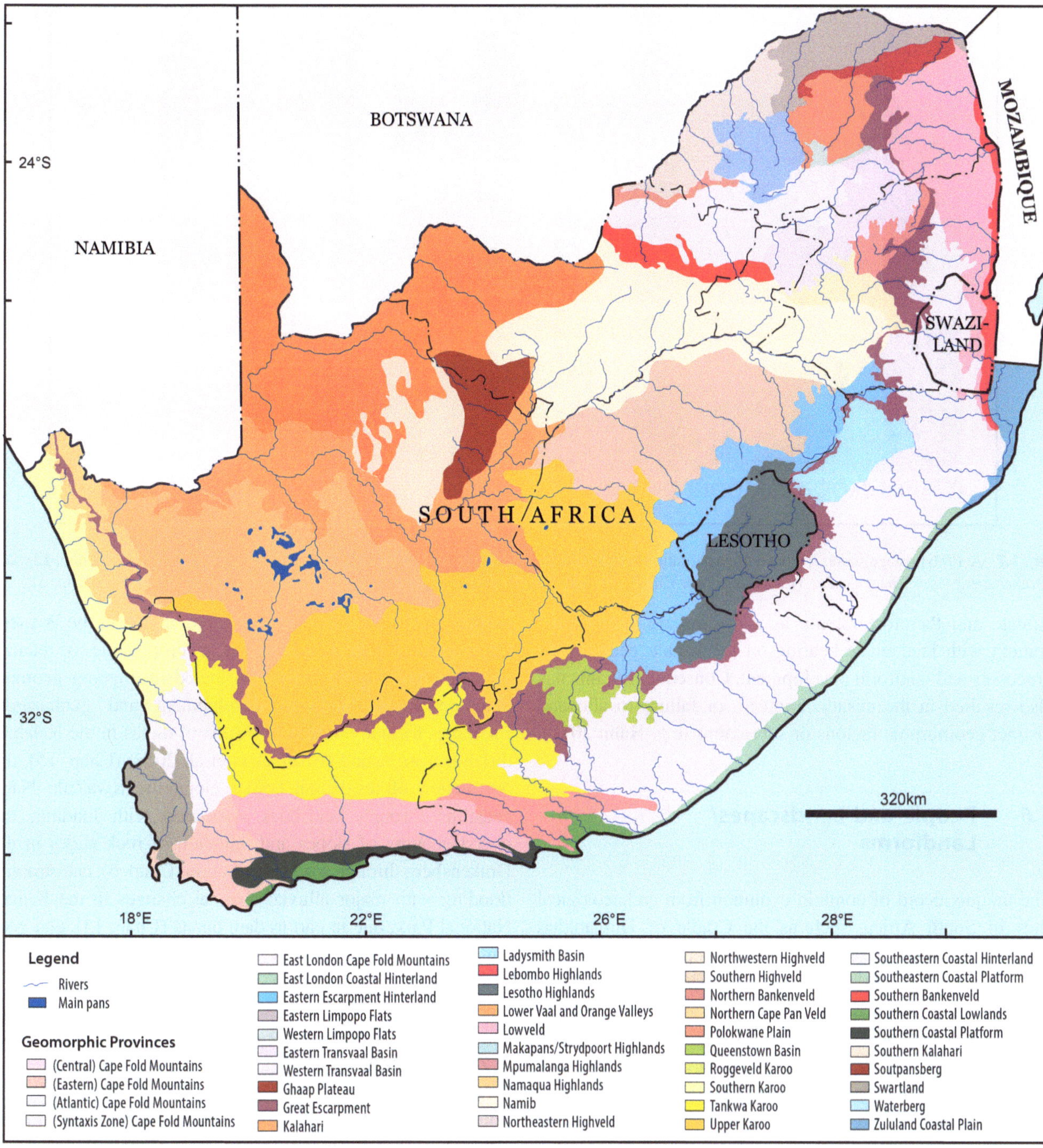

Fig. 1.6 The geomorphic provinces of South Africa (redrafted after Partridge et al. 2010)

preservation of palaeo-landforms such as Late Quaternary glacial deposits (Chap. 6), raised and contemporary coastal shore platforms (Chap. 7), raised Pliocene shorelines and ancient coastal lakes and dunes (Chap. 14), arid zone pans, dunes and palaeo-drainage systems (Chaps. 9, 15 and 16), and subterraneous karstic deposits (Chaps. 3 and 17), to name a few. Using a hierarchy of criteria including geomorphic history, geological structure, climate, location and altitude, Lester King originally delineated 26 geomorphic provinces for southern Africa, but later refined these to 18 (King 1942, 1967). More recently, Partridge et al. (2010) identified 34 geomorphic provinces and 12 subprovinces based on new digital terrain model (DTM)-derived data and statistical techniques which placed a strong focus on drainage structure and slope (Fig. 1.6). These classifications, however, are unable to adequately incorporate local-scale

Fig. 1.7 A 1986 postage stamp special issue celebrating South African landscapes and landforms: *14c* (Paarlberg, Winelands—see Chap. 12); *20c* (Drakensberg—see Chap. 6); *25c* (Cederberg—see Chap. 10); and *30c* (Bourke's Luck potholes, Mpumalanga Escarpment—see Chap. 3)

climate and therefore remain as a macroscale geomorphic context with inadequate bearing on local-scale geomorphic processes and landform development. Consequently, this has also resulted in the misallocation of, or failure to allocate, distinct geomorphic regions or subregions (e.g. Hahn 2011).

1.6 People and Landscapes/Landforms

The unique record of hominin evolution from archaeological sites in South Africa, such as the Cradle of Humankind (Chap. 17), shows the intimate relationship between climate and environmental changes over long timescales, and associated patterns of human development (Vrba et al. 1995). Several rock shelters occupied by hominids since ~80 Ka at places such as Diepkloof, Western Cape (Tribolo et al. 2009), and Sibudu, KwaZulu–Natal (Wadley et al. 2011), attest to a long association between people and South African landscapes and landforms. More recently, the San ('Bushman') left their legacy of rock art on the walls of many caves and rock overhangs (Chaps. 2 and 10) and early settlers strategically used the landscape, such as during the Anglo-Boer war (Chap. 18). Many of today's South African cities and towns were carefully located, utilizing the landscape for advantage, such as alongside perennial river systems in otherwise near-uninhabitable drylands (e.g. Kuruman, Upington), places with safe harbours (e.g. Cape Town; Chap. 11) or for river-port development (e.g. East London, Richards Bay). Landscapes, or more particularly slope positions with associated soil characteristics and microclimates, are additionally exploited to maximize agricultural outputs, such as for fruit and wine production in the Western Cape region (Chap. 12). However, anthropogenic (or related) activities have significantly contributed to ongoing geomorphic processes, landform development and geohazards. Examples include the remobilization of dunes in the Kalahari through overgrazing (Bhattachan et al. 2014) (Chap. 15), the destruction of residential homes along the KwaZulu–Natal coastline through landslides associated with loading, the oversteepening of slopes and initiation of rock slides in the Drakensberg through road construction (Chap. 6), catastrophic flooding with major alluvial channel changes in the Kruger National Park, due in part to dam bursts (Chap. 13), and geovandalism (e.g. graffiti over San rock art or on geomorphic landforms of interest) (Chap. 19).

1.7 Conclusions

South Africa undoubtedly has a rich geomorphological heritage which has been recognized and celebrated through a variety of mechanisms such as the display of iconic landscapes and landforms on postage stamps (Fig. 1.7), posters at ports of entry, national parks and reserves, and on tourist websites. Tourism is currently a major foreign income-earning industry in South Africa, with a large drawcard being its landscapes. Indeed, many landscapes or geologic/geomorphic phenomena across South Africa are attributes that form part of provincial or national heritage sites, being inscribed as World Heritage Sites (e.g. Vredefort Dome, Chap. 4; Drakensberg, Chap. 6; Richtersveld, Chap. 9; iSimangaliso Wetlands, Chap. 14; Sterkfontein Caves, Chap. 17), or listed among the Wonders of

Nature (e.g. Table Mountain, Chap. 11) (see also Chap. 19). The wider context for landscape management and conservation in South Africa is the backdrop of ongoing climate change, biodiversity loss, increased mining and urbanization, increased water scarcity, and increased need for agricultural production. These factors in combination are exerting considerable pressure on the integrity of whole landscapes as well as specific geomorphological sites, and thus, future planning and management requires good scientific information as well as an appreciation of geomorphic resources.

References

Beckedahl HR, Sumner PD, Garland G (2002) South African geomorphology: Critical choices for the future. S Afr Geogr J 84:145–152

Bhattachan A, D'Odorico P, Dintwe K, Olin GS, Collins SL (2014) Resilience and recovery potential of denuded vegetation in the southern Kalahari. Ecosphere 5:1–14

Bierman PR, Caffee M (2001) Slow rates of rock surface erosion and sediment production across the Namib Desert and escarpment, southern Africa. Am J Sci 301:326–358

Bishop P (2007) Long-term landscape evolution: linking tectonics and surface processes. Earth Surf Proc Land 32:329–365

Brown RW, Summerfield MA, Gleadow AJW (2002) Denudational history along a transect across the Drakensberg Escarpment of southern Africa derived from apatite fission track thermochronology. J Geophys Res (Solid Earth) 107:74–77

Burke K, Steinberger B, Torsvik TT, Smethurst MA (2008) Plume generation zones at the margins of large low shear velocity provinces on the core-mantle boundary. Earth Planet Sci Lett 286:49–60

Button A, Vos RG (1977) Subtidal and intertidal clastic and carbonate sedimentation in a macrotidal environment: an example from the lower Proterozoic of South Africa. Sed Geol 18:175–200

Eales HV, Marsh JS, Cox KG (1984) The Karoo igneous province: an introduction. In: Erlank AJ (ed) Petrogenesis of the Volcanic Rocks of the Karoo Province (special publication No. 3). Geological Society of South Africa, Johannesburg

Eriksson PG, Reczko BFF, Boshoff AJ, Schreiber UM, Van der Neut M, Snyman CP (1995) Architectural elements from Lower Proterozoic braid-delta and high-energy tidal flat deposits in the Magaliesberg Fm, Transvaal Supergroup, South Africa. Sed Geol 97:99–117

Flowers RM, Schoene B (2010) (U-Th)/He thermochronometry constraints on unroofing of the eastern Kaapvaal craton and significance for uplift of the southern African plateau. Geology 38:827–830

Fouché J, Bate KJ, van der Merwe R (1992) Plate tectonic setting of the Mesozoic Basins, southern offshore, South Africa: a review. In: de Wit MJ, Ransome IGD (eds) Inversion tectonics of the cape fold belt, Karoo and Cretaceous Basins of Southern Africa. Balkema, Rotterdam, pp 33–45

Gilchrist AR, Summerfield MA (1990) Differential denudation and flexural isostasy in formation of rifted-margin upwards. Nature 346:739–742

Goudie AS, Wells GL (1995) The nature, distribution and formation of pans in arid zones. Earth Sci Rev 38:1–69

Hahn N (2011) Refinement of the Soutpansberg Geomorphic Province, Limpopo, South Africa. Trans R Soc S Afr 66:32–40

Holmes P, Meadows M (eds) (2012) Southern African geomorphology: recent trends and new directions. Sun Press, Bloemfontein

Isbell JL, Cole DI, Catuneanu O (2008) Carboniferous-Permian glaciations in the main Karoo Basin, South Africa: stratigraphy, depositional controls, and glacial dynamics. Geol Soc Am Spec Pap 441:71–82

Japsen P, Chalmers JA, Green PF, Bonow JM (2012) Elevated, passive continental margins: not rift shoulders, but expressions of episodic, post-rift burial and exhumation. Global Planet Change 90–91:73–86

King LC (1942) South African scenery. Oliver and Boyd, Edinburgh

King LC (1944) Geomorphology of the Natal Drakensberg. Trans Geol Soc S Afr 47:255–282

King LC (1955) Pediplanation and isostasy: an example from South Africa. Q J Geol Soc London 111:535–539

King LC (1967) The morphology of the Earth, 2nd edn. Oliver and Boyd, Edinburgh

Moon BP, Dardis GF (eds) (1988) The geomorphology of southern Africa. Southern Book Publishers, Johannesburg

Moore A, Larkin P (2001) Drainage evolution in south-central Africa since the breakup of Gondwana. S Afr J Geol 104:47–68

Moore A, Blenkinsop T, Cotterill F (2009) Southern African topography and erosion history: plumes or plate tectonics? Terra Nova 21:310–315

Noffke N, Eriksson KA, Hazen RM, Simpson EL (2006) A new window into Early Archean life: microbial mats in Earth's oldest siliclastic tidal deposits (3.2 Ga Moodies group, South Africa). Geology 34:253–256

Partridge TC (1997) Late Neogene uplift in eastern and southern Africa and its palaeoclimatic implications. In: Ruddiman WF (ed) Tectonic uplift and climate change. Plenum Press, New York, pp 63–86

Partridge TC (1998) Of diamonds, dinosaurs and diastrophism: 150 million years of landscape evolution in southern Africa. S Afr J Geol 101:167–184

Partridge TC, Maud RR (2000a) The Cenozoic of southern Africa. Oxford University Press, Oxford 406 pp

Partridge TC, Maud RR (2000b) Macro-scale geomorphic evolution of southern Africa. In: Partridge TC, Maud RR (eds) The Cenozoic of southern Africa. Oxford University Press, Oxford, pp 3–18

Partridge TC, Dollar ESJ, Moolman J, Dollar LH (2010) The geomorphic provinces of South Africa, Lesotho and Swaziland: a physiographic subdivision for earth and environmental scientists. Trans R Soc S Afr 65:1–47

Rosenthal E (1970) Gold! Gold! Gold!: the Johannesburg Gold Rush. Macmillan, London

Smith RMH (1995) Changing fluvial environments across the Permian-Triassic boundary in the Karoo Basin, South Africa and possible causes of tetrapod extinctions. Palaeogeogr Palaeoclimatol Palaeoecol 117:81–104

Tinker J, de Wit M, Brown R (2008) Mesozoic exhumation of the southern Cape, South Africa, quantified using apatite fission track thermochronology. Tectonophysics 455:77–93

Tribolo C, Mercier N, Valladas H, Joron JL, Guilbert P, Lefrais Y, Selo M, Texier P-J, Rigaud J-Ph, Porraz G, Poggenpoel C, Parkington J, Texier J-P, Lenoble A (2009) Thermoluminescence dating of a Stillbay-Howiesons Poort sequence at Diepkloof Rock Shelter (Western Cape, South Africa). J Archaeol Sci 36:730–739

Vrba ES, Denton GH, Partridge TC, Burckle LH (eds) (1995) Paleoclimate and evolution, with emphasis on human origins. Yale University Press, New Haven

Wadley L, Sievers C, Bamford M, Goldberg P, Berna F, Miller C (2011) Middle Stone Age bedding construction and settlement patterns at Sibudu, South Africa. Science 334:1388–1391

Young GM, von Brunn V, Gold DJC, Minter WEL (1998) Earth's oldest reported glaciation: physical and chemical evidence from the Archean Mozaan Group (~2.9 Ga) of South Africa. J Geol 106:523–538

References

Sandstone Landforms of the Karoo Basin: Naturally Sculpted Rock

2

Stefan Grab

Abstract

South Africa boasts some of the most impressive sandstone landscapes and landforms in the world, and although these are widely distributed across South Africa, some of the most spectacular examples are associated with the Molteno, Elliot and Clarens Formations in the central region of South Africa. The prominence of sandstone in this region is primarily owing to palaeo-basin infilling during the late Carboniferous and a climate dominated by seasonal precipitation patterns, both now and in the past. Consequently, a range of weathering and erosion processes have operated at wide-ranging spatial scales upon the sandstone outcrops. The chapter describes prominent sandstone landscapes (plateaus, mesa-butte topography, scarplands, slopes) and landforms (e.g. ichnofossil structures, honeycombs, rock arches, rock doughnuts) of central South Africa and reflects on their associated cultural heritage and geoheritage linkages. For instance, almost all known San rock art sites are associated with sandstone, yet rapid weathering of such rock is jeopardizing the longevity of this cultural legacy.

Keywords

Sandstone • Karoo basin • Weathering • Cultural heritage/geoheritage

2.1 Introduction

A considerable portion of South Africa's landscape is characterized by sedimentary rocks which represent infilled basins such as the Main Karoo basin, which is now the primary host to the subcontinent's sandstone landscapes and landforms (Fig. 2.1). Importantly, this basin contains rich palaeontological and hominid fossil records (e.g. Sterkfontein), and world-renowned rock art sites (e.g. Kamberg, Main Caves at Giant's Castle), which therefore has considerable geotourism appeal. It is likely that the spectacular sandstone landscape and landforms also fascinated early settlers, as attested by place names such as 'Kamieskroon' (kroon = 'crown', to describe the prominent sandstone rock formation upon a hill) in the Western Cape, and 'Maanhaarrand' (meaning 'mane of a horse' which describes a sandstone ridge outcrop) in the Magaliesberg (Fig. 2.1).

Early sandstone work in South Africa was primarily concerned with describing the macroscale landscape of slope forms (e.g. King 1953, 1957; Le Roux 1978; Moon and Munro-Perry 1988). In contrast, more recent work has described the rich variety of sandstone landforms in areas such as the eastern Free State Province (Grab et al. 2011), suggested process mechanisms for rock doughnut formation (Grab and Svensen 2011), and undertaken experimental work on sandstone weathering (e.g. Mol 2014). This chapter provides the geological context to sandstone landscapes in the central Karoo Basin region of South Africa including areas in the Free State and Eastern Cape and illustrates examples of some of the more intriguing landforms found within them. Finally, the strong connectivity between sandstone geomorphology and cultural (stone) heritage

S. Grab (✉)
University of the Witwatersrand, Johannesburg, South Africa
e-mail: stefan.grab@wits.ac.za

S. Grab and J. Knight (eds.), *Landscapes and Landforms of South Africa*,
World Geomorphological Landscapes, DOI 10.1007/978-3-319-03560-4_2,

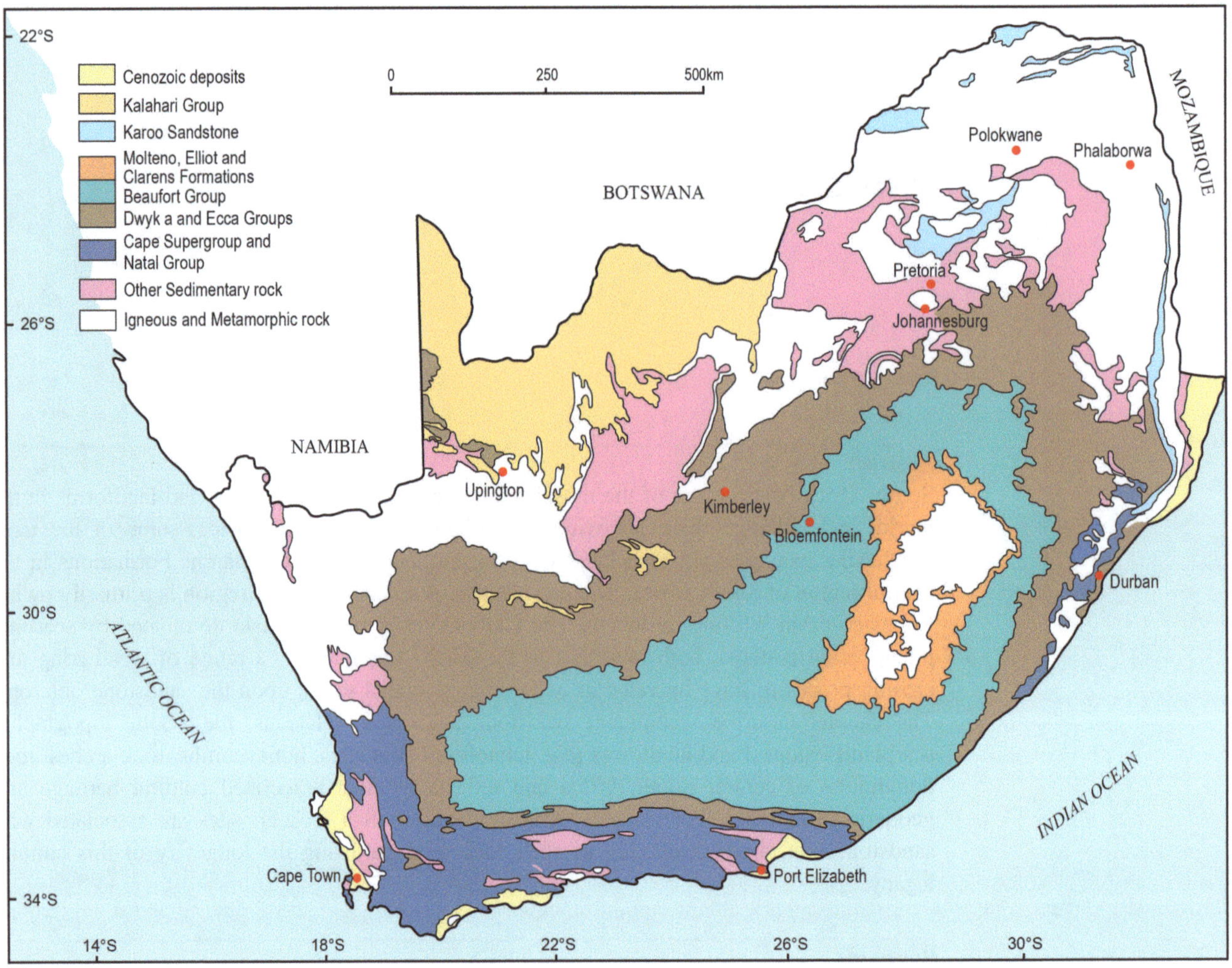

Fig. 2.1 Sandstone/sedimentary outcrops in South Africa, Lesotho and Swaziland (redrafted after Council for Geoscience, 1997; Stefan Grab and Wendy Phillips)

provides for mixed (i.e. natural and cultural) heritage tourist attractions, for which these regions are already well known. Places referred to in the text are represented in Fig. 2.2.

2.2 Geological Setting

Most sedimentary rocks in South Africa are a product of basin infills, of which there are several including the Main Karoo Basin, which existed over much of the central interior of South Africa (Fig. 2.1). This chapter will focus on this region specifically.

The sandstone succession in the Karoo Basin accumulated over a period spanning from the late Carboniferous (~300 Ma) to the early Jurassic (~190 Ma). Climate change over this time period influenced sediment type, depositional environments and geological structures (e.g. bedding planes) of the various sandstone formations. In turn, these have influenced the spatially variable weathering and erosion processes over time. Subsequent infilling of the basin yielded a total stratigraphic thickness of ~12 km, and the Karoo Supergroup currently crops out over more than half of South Africa (~600,000 km^2). However, the most striking sandstone landscapes/landforms are those associated with the Molteno, Elliot and Clarens Formations of the Karoo Supergroup. The late Triassic Molteno Formation outcrops over ~25,000 km^2, consists of rock types ranging from mudstones to coarse sandstones, and attains a maximum thickness of ~460 m (Turner 1983). The Elliot Formation (formerly known as the 'red beds') represents a transition from predominantly fluvial and lacustrine during the late Triassic to increasingly aeolian during the early Jurassic (Lucas and Hancox 2001) (Fig. 2.2). The 'Lower Elliot Formation' consists of thick mudstones and isolated

Fig. 2.2 Locations of sandstone phenomena and places mentioned in the text of this chapter

channel sandstones, while the fine- to medium-grained and interbedded mudstones of the 'Upper Elliot Formation' are thought to be the product of an ephemeral flash flood-dominated fluvial system (Bordy et al. 2004a).

The final phase of basin infill was associated with shallowing-upward lake infill with alluvial fan propagation on the northern basin margin (Eriksson 1986; Holzförster 2007). The resulting Clarens Formation (Fig. 2.3) was previously known as 'cave sandstone formation' owing to the prominent concavities/overhangs at the base of near-vertical cliffs. Although the Clarens Formation is represented by the finest textured sandstones within the sandstone stratigraphic sequence, a variety of sedimentary rock types ranging from mudstones and shale to coarse conglomeratic sandstones has been identified within it.

2.3 Sandstone Landscapes

The sandstone landscapes of central South Africa have an important place in global geomorphology, as this is where Lester King (1953, 1957) developed his ideas of parallel slope retreat. More recently, Moon and Munro-Perry (1988) tested the concept of parallel slope retreat in the sandstone landscapes of the north-eastern Free State province; they found a variety of sandstone slope forms and concluded that many site-specific conditions determine hillslope development. In most instances, valley side slopes develop by cliff retreat and subsequent replacement by rectilinear bedrock slopes, which may at first be mantled by weathered debris. The sandstone slope forms are well preserved over geological

Fig. 2.3 The Elliot Formation ('red bed') underlying the aeolian Clarens Formation, eastern Free State (the *arrow* indicates the distinct boundary between the two formations) (*Photograph* S. Grab)

time scales owing to the relatively resistant nature of the sandstone bedrock (Moon and Munro-Perry 1988).

The western plateau regions, especially in the eastern Free State, are the product of long periods of landscape denudation associated with predominantly parallel scarp retreat through large drainage meander systems undercutting slopes and causing cliffline failure and colluviation. Consequently, this is a landscape of mesas and buttes (Fig. 2.4); here, the cliff faces are relatively smooth and uniform, suggesting little contemporary spatial variation in rates of cliffline retreat. The typical succession of convex interfluves—near-vertical cliffs—linear colluvial mid-slopes—and concave foot slopes is notable in this landscape. In the wetter north-eastern Free State, the lower slopes are relatively stable and vegetated, but as one travels towards the south-western Karoo Basin, the drier climate exposes lower slope elements to more visible contemporary denudation (active gullies and colluvial fans). The mesa-butte sandstone landscape is primarily controlled by a more resistant upper basalt or Clarens sandstone outcrop, which in places provides a distinct lateral protrusion or capping over the underlying light yellowish Clarens and reddish Elliot sandstones (Fig. 2.3). The stratigraphic positioning of more highly jointed lower Clarens and Elliot sandstones (often represented as mudstones) regularly coincides with areas of groundwater seepage (sapping), hence creating the conspicuous overhangs that contour slopes for hundreds of metres. Such undercutting also causes widespread block/toppling failure, which is still currently active and contributes to rock fall debris mantles.

By contrast, the sandstone landscape to the east of the Great Escarpment is a product of rapid incision enhanced by a dense drainage network and wet climate. Upper catchments are within the Drakensberg basalts and are largely controlled by major basaltic lineaments and dolerite intrusions. The drainage network produced relatively deeply incised sandstone valleys (with narrow gorges, waterfalls and cascades) in the upper reaches (Fig. 2.5a), yet elsewhere ridges and valleys are uniformly spaced and asymmetric in cross-profile. Valley-side drainage off sandstone interfluves in several places has also initiated very regular patterns of bedrock 'ribs' and adjacent eroded channels, likely controlled by bedrock mechanical strength and climate.

On interfluves, landforms include 'tors' or 'domes' and 'pillars' (Fig. 2.5b), which in some instances have become tourist attractions named after the rock formation, such as the 'Policeman's helmet' at Royal Natal National Park or

Fig. 2.4 A butte in the eastern Free State, west of the Great Escarpment (*Photograph* S. Grab)

Fig. 2.5 **a** Steep, incised sandstone valley to the east of the Great Escarpment, Drakensberg foothills; **b** mushroom shaped 'tor'/'pillar'/'dome', Sehlabathebe National Park, Drakensberg (~17 m in height) (*Photographs* S. Grab)

'Mushroom rock' outside the town of Clarens. These are thought to be a product of initial subsurface etching, and once exposed are subject to enhanced weathering and erosion associated with joints and areas of lower rock strength at the base of these phenomena (Ollier 1978).

2.4 Sandstone Landforms

More striking than sandstone landscapes is the diversity of specific sandstone landforms within them, which vary from micro-scale (mm) to tens of metres in size. These are primarily a product of site-specific weathering and erosion over geological and shorter timescales. Several conspicuous contemporary sandstone landforms are also a product of geohydrological or biological activity of the past. Here, a few of the most striking landforms are described.

Pentagonally or hexagonally *fractured sandstone surfaces* (crack diameters = ~2–5 cm; surface diameters = ~5 to >50 cm) are particularly common in case-hardened sandstone outcrops (Fig. 2.6a). Suggested mechanisms include shrinkage of silica gel due to changing rock thermal and/or moisture conditions (Robinson and Williams 1992) and thermal surface stress (Croll 2009), although it remains

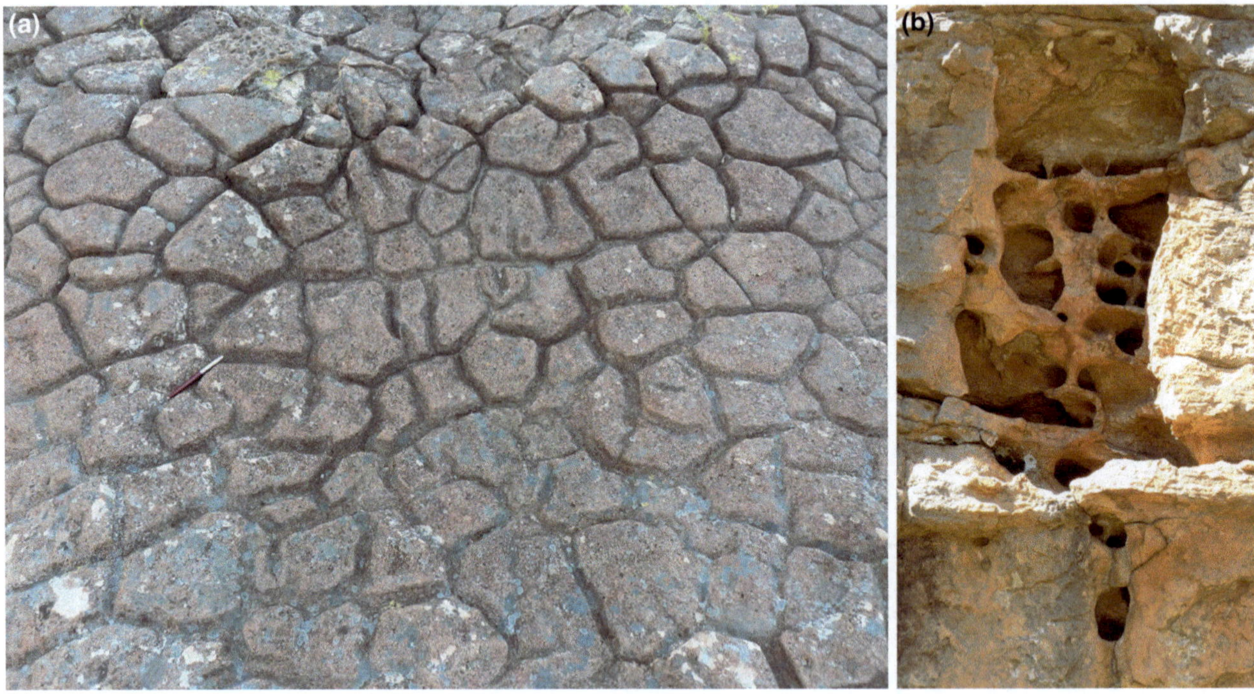

Fig. 2.6 **a** Fractured sandstone surface on case-hardened rock, eastern Free State (pen for scale); **b** honeycombs, in which both birds and rodents build their nests, Eastern Cape (total height = ~220 cm) (*Photographs* S. Grab)

unclear whether these are syndepositional or postdepositional phenomena. In many instances, such fractures direct surface run-off producing rinnenkarren (rundkarren) or rillenkarren ('solution flutes'), which are likely strongly associated with dissolution and biokarstic processes. In some instances, long micro-drainage channels connect a succession of shallow rock basins.

Honeycombs, consisting of closely spaced pits bounded by thin raised lips or walls (McBride and Picard 2004), are widespread in sandstone outcrops throughout South Africa (Fig. 2.6b). Such micro-weathering forms (few tens of cm across), also known as 'alveoli', are commonly ascribed to salt solutional and salt recrystallization processes. In contrast, case-hardening by iron oxide or calcium carbonate provides greater resistance to weathering between the pits (McBride and Picard 2004). Somewhat larger *tafoni* (several metres across) cavernous weathering/erosion forms also occur, albeit less frequently. These typically have overhanging lips/hoods/visors, arch-shaped entrances, concave inner walls, overhanging margins and relatively smooth, gently sloping, debris-covered floors (Grab et al. 2011).

The largest cavernous sandstone forms are those commonly referred to as *amphitheatres* or *alcoves* (e.g. Laity and Malin 1985), which overhang below the cliffline. These may be up to several hundred metres in diameter and over 100 m in height. Several mechanisms may be responsible; these include groundwater sapping and perched water tables locally weakening rock, granular disintegration, wind erosion, rock falls and plunge pool scour at the base of waterfalls, all of which produce caverns below vertical cliffs. These processes are most enhanced where weaker, less permeable sedimentary strata underlie stronger or case-hardened cap rock (Ryan et al. 2012).

Apart from coastal *rock arches* (see Chap. 7), sandstone arches are best known from the Cederberg ('Wolfberg Arch': see Chap. 10) and southern Drakensberg/Lesotho regions of South Africa, with smaller examples also occurring across the Karoo, Eastern Cape, Magaliesberg and Mpumalanga escarpment regions. Natural bridges, which are rock arches that span across valleys, have been documented during historical times from the Golden Gate Highlands National Park (Fig. 2.7), two in the Mpumalanga Province (see Chap. 19) and one in the remote Lesobeng Valley in central Lesotho—which is also likely the largest rock arch in southern Africa at ~70 m across, ~9 m in height and ~4 m in bedrock thickness at the centre of the arch. Arches, in contrast to natural bridges, do not span valleys but are typically found in canyon lands (Dixon 2010) and denuded sandstone-dominated plateaus and interfluves, such as is the case at Sehlabathebe National Park (SNP) (Fig. 2.8). Here is found the highest concentration of rock arches in southern Africa (~10–12); these are typically 5–12 m in diameter and ~4–5 m in height. Most sandstone rock arches are thought to be the product of enhanced weathering and erosion along joint/fracture zones,

Fig. 2.7 A natural rock bridge (now collapsed) in the Golden Gate Highlands National Park, as depicted in a photograph by Johan Van Reenen, dating to ca 1930–1940 (*Source* South African National Parks)

Fig. 2.8 Rock arch in Clarens Fm. sandstone, Sehlabathebe National Park, Drakensberg (~7 m in width; ~5 m in height) (*Photograph* S. Grab)

often associated with rockfalls along horizontal or vertical discontinuities, eventually leading to material collapse and production of the 'window area' (Leasure et al. 2007).

A variety of raised conical- or cone-shaped structures, often with a central pothole (rock basin), have been described from several Clarens Formation sandstone outcrops in

South Africa, including the Eastern Cape, KwaZulu-Natal, Free State and Limpopo provinces. These are typically known as *rock doughnuts*, which are circular/oval in plan, and many are bowl shaped in cross section with a pothole in the centre. Potholes develop singly or in clusters, are flat floored and contain weathered detritus. Rock basins may contain standing water for short periods after rainfall events, or for several weeks to months during wet periods/seasons, and induce both chemical weathering associated with dissolution (Domínguez-Villar et al. 2009), and mechanical (Goudie and Migoń 1997) weathering.

The hypotheses for rock doughnut formation range from differential water-level weathering/erosion due to water/moisture contrasts near the surface (e.g. Twidale and Vidal Romaní 2005) to lithological variations, liquefaction and fluidization, which produce clastic pipes (Netoff and Shroba 2001; Grab and Svensen 2011). In places, small and circular pipe structures filled with calcite-cemented sand are present and suggest conduits for fluidized sand. This may suggest that some of these features are Jurassic remnants cemented with superimposed and secondary silica (Grab and Svensen 2011).

2.5 Biological Influences on Sandstone Formations

Biota such as lichens, fungi and cyanobacteria are known to shape sandstone surfaces, causing relatively small-scale (mm–cm) etching, pitting and flaking (see Viles et al. 2008). Larger-scale (cm–m) fossilized sandstone phenomena (ichnofossils) in the form of insectivorous nests (*calie*) have also been reported from central South Africa.

Most sandstone surfaces in central South Africa are at least partially lichen covered and are thus weathered through processes including cation chelation, dissolution, swelling and hyphal penetration (see Grab et al. 2011). In addition, the larvae of a bagworm moth species (*Lepidoptera*) associated with endolithic lichen further contribute to weathering by dissolving the sandstone-cementing agents (Wessels and Wessels 1991). Cryptoendolithic cyanobacteria (blue-green algae) are also widespread, and typically so along sandstone fractures where weathering is induced by substrate alkalization during photosynthesis (Büdel et al. 2004). These processes are responsible for micro-weathering forms such as circular or crescentric-shaped depressions bounded by micro-scarps (Fig. 2.9a). The sandstone rock basins or potholes mentioned in Sect. 2.4 are not only a product of mechanical geomorphic processes, but also closely connected to cyanobacteria and other micro-organisms found in the microbial mats within sandstone depressions. Complex biofilms may accumulate at the base of rock basins, dissolving cementing agents between sandstone grains, but also act as biological sealants to water infiltration (see Chan et al. 2005).

Unique *sandstone pillars* (Fig. 2.9b) up to 3.3 m in height, possibly representing fossil termite (or other social soil-dwelling organism) nests, are prominent in the eastern Free State/western Lesotho and Tuli Basin regions (Bordy et al. 2004b, 2009). These ichnofossils occur in groups, typically spaced between 0.5 and 10 m. The true height of the features is difficult to determine owing to denudation at the top of the pillars and extension into the parent bedrock at the base, but may typically stand 1–3 m above ground. The basal circumference commonly exceeds 1 m, but

Fig. 2.9 **a** A lichen-weathered sandstone surface: note the crescentric-shaped depressions bounded by micro-scarps (*Photograph* S. Grab); **b** sandstone ichnofossils of likely termite origin, Limpopo Province (*Photograph* E. Bordy)

pillars narrow towards their apices. Although the outer walls are relatively smooth, the internal bioturbated structure consists of interconnected burrows, channels, ducts and tubes averaging 0.5–2 cm in diameter. The pillars are exposed due to differential rates of weathering and erosion between the nest sites and surrounding sandstone, even though both are of aeolian origin. The elongated north–south-orientated nests are thought to be a bioengineered midday heat deflection construction in what would have been an arid hot environment (Bordy et al. 2004b). The biochemical process involved produces a hydrophobic film, which, together with the general absence of swelling clays but abundance of quartz minerals in Clarens Formation sandstones, lowers shrinkage but increases the bonding mechanism. Consequently, surrounding sandstone would have weathered and eroded more rapidly, thus exposing and preserving the ichno-fossiliferous sandstone pillars.

2.6 Sandstone Geomorphology and Cultural Heritage

There are exceptionally strong associations between southern African sandstone and cultural heritage/geoheritage. For instance, the SNP was added to the World Heritage Site List (mixed cultural/natural category) in 2013, in part given its sandstone landscape appeal. Sandstone overhangs (*alcoves*) were important occupation sites by the San ('Bushman') for at least the last two thousand years. In addition, ancient rock dwellings from former hunter-gatherer and early farmers still occur in some sandstone overhangs (Fig. 2.10a). There were even accounts of missionary workers during the nineteenth century taking refuge in such rock shelters during times of war. Elsewhere, sandstone potholes have been used in former times for fire making, and surfaces used for engraving outlines for the Morabaraba

Fig. 2.10 Sandstone cultural heritage: **a** rock dwelling within a sandstone cavity (indicated by *red arrow*), **b** the morabaraba board game (note the lichen growth along the engraved sections, indicating pronounced micro-spatial bioweathering) and **c** weathered San rock art (*Photographs* S. Grab)

board game (Lesotho variant of Nine Men's Morris board game) (Fig. 2.10b).

Arguably the best example of mixed cultural/sandstone heritage is that through the preservation of San rock art on sandstone surfaces (Fig. 2.10c). Several thousand rock art sites are known, almost all of which are located within finer-grained sandstone outcrops (particularly Clarens Formation), owing to their smooth and light-coloured surfaces. Rock art sites frequently occur within overhangs where active weathering makes the paintings fragile. Water seepage and thermal stress at such sites have been identified as contributing factors in their disintegration (e.g. Hoerlé 2006). Thus, considerable attention has been given to understanding rock weathering and art deterioration at key field sites (e.g. Mol and Viles 2010). However, protecting such art in the natural environment is exceptionally difficult. One option is to identify the most critical rock art panels likely to detach in the near future, through techniques such as GPR profiling (e.g. Denis et al. 2009), and possibly to retrieve these panels from such sites—although this action is controversial.

2.7 Concluding Remarks

Sandstone weathering and erosion has produced remarkable landscapes, landforms and niches for cultural preservation in the central Karoo Basin of South Africa; arguably one of the best examples globally. It was through the observation and theorization of South African sandstone landscapes by the likes of King that brought attention to southern Africa's rich geomorphological heritage. In contrast, more recent micro-scale approaches towards understanding sandstone geomorphic processes, particularly in the context of weathering associated with rock art deterioration/preservation, have highlighted the global value of this natural geomorphic laboratory. The longevity and preservation of some sandstone landforms and associated cultural artefacts are now threatened through climate change, inappropriate management, land transformation and tourism, and thus require better attention in future.

Acknowledgments I thank Lisa Mol, Piotr Migoń, John Dixon and Jasper Knight who provided constructive comments on an earlier draft—their input has surely improved the quality of this chapter.

References

Bordy EM, Hancox PJ, Rubidge BS (2004a) A description of the sedimentology and palaeontology of the late Triassic-early Jurassic Elliot Formation in Lesotho. Palaeoentologia Africana 40:43–58

Bordy EM, Hancox PJ, Rubidge BS (2004b) Fluvial style variations in the late triassic-early Jurassic Elliot Formation, main Karoo Basin, South Africa. J Afr Earth Sci 38:383–400

Bordy EM, Bumby AJ, Catuneanu O (2009) Possible trace fossils of putative Jurassic (Karoo supergroup) of South Africa and Lesotho. S Afr J Sci 105:356–362

Büdel B, Weber B, Kühl M, Pfanz H, Sültemeyer D, Wessels D (2004) Reshaping of sandstone surfaces by cryptoendolithic cyanobacteria: bioalkalization causes chemical weathering in arid landscapes. Geobiology 2:261–268

Chan MA, Moser K, Davis JM, Southam G, Hughes K, Graham T (2005) Desert potholes: ephemeral aquatic microsystems. Aquat Geochem 11:279–302

Croll JGA (2009) Possible role of thermal ratcheting in alligator cracking of asphalt pavements. Int J Pavement Eng 10:447–453

Denis A, Huneau F, Hoerlé S, Salomon A (2009) GPR data processing for fractures and flakes detection in sandstone. J Appl Geophys 68:282–288

Dixon JC (2010) Canyonlands and arches: windows on landscapes in the American southwest. In: Migoń P (ed) Geomorphological landscapes of the world. Springer, Dordrecht. 39–47

Domínguez-Villar D, Razola L, Carrasco RM, Jennings CE, Pedraza J (2009) Weathering phases recorded by gnammas developed since last glaciation at Serra da Estrela, Portugal. Quatern Res 72:218–228

Eriksson PG (1986) Aeolian dune and alluvial fan deposits in the Clarens Formation of the Natal Drakensberg. Trans Geol Soc S Afr 89:389–394

Goudie AS, Migoń P (1997) Weathering pits in the Spitzkoppe area, central Namib desert. Z Geomorphol 41:417–444

Grab SW, Svensen H (2011) Rock doughnut and pothole structures of the Clarens Fm. Sandstone in the Karoo Basin, South Africa: possible links to Lower Jurassic fluid seepage. Geomorphology 131:14–27

Grab SW, Goudie AS, Viles HA, Webb N (2011) Sandstone geomorphology of the Golden Gate Highlands National Park, South Africa, in a global context. Koedoe 53: Art. #985, 14 p. doi:10.4102/koedoe

Hoerlé S (2006) Rock temperatures as an indicator of weathering processes affecting rock-art. Earth Surf Proc Land 31:383–389

Holzförster F (2007) Lithology and depositional environments of Lower Jurassic Clarens Formation in the Eastern Cape, South Africa. S Afr J Geol 110:543–560

King LC (1953) Canons of landscape evolution. Bull Geol Soc Am 64:721–751

King LC (1957) The uniformitarian nature of hillslopes. Trans Edinb Geol Soc 17:81–102

Laity JE, Malin MC (1985) Sapping processes and the development of theater-headed valley networks on the Colorado plateau. Geol Soc Am Bull 96:203–217

Le Roux JS (1978) Die ontstaan van grotte en holtes in die kranse van die Clarenssandsteenformasie in die Golden Gate-gebied [Formation of caves and hollows in the cliffs of the Clarens Sandstone Formation in the Golden Gate area]. S Afr Geogr 17:129–131

Leasure VL, Kind TC, Busby MR (2007) Formation of Apex natural arch, Kentucky. J Ky Acad Sci 68:96–101

Lucas SG, Hancox PJ (2001) Tetrapod-based correlation of the nonmarine Upper Triassic of Southern Africa. Albertiana 25:5–9

McBride EF, Picard MD (2004) Origin of honeycombs and related weathering forms in Oligocene Macigno Sandstone, Tuscan coast near Livorno, Italy. Earth Surf Proc Land 29:713–735

Mol L (2014) Investigations into the relationship between changes in internal moisture regimes and rock surface deterioration in cavernous sandstone features. Earth Surf Proc Land 39:914–927

Mol L, Viles HA (2010) Geoelectric investigations into sandstone moisture regimes: implications for rock weathering and the deterioration of San rock art in the Golden Gate Reserve, South Africa. Geomorphology 118:280–287

Moon BP, Munro-Perry PM (1988) Slope development on the Clarens Sandstone Formation in the northeastern Orange Free State. S Afr Geogr J 70:57–68

Netoff DI, Shroba RR (2001) Conical sandstone landforms cored with clastic pipes in Glen Canyon National Recreation Area, southeastern Utah. Geomorphology 39:99–110

Ollier CD (1978) Induced fracture and granite landforms. Zeitschrift für Geomorphologie 22:249–257

Robinson DA, Williams RBG (1992) Sandstone weathering in the High Atlas, Morocco. Z Geomorphol 36:413–429

Ryan AJ, Whipple KX, Johnson JP (2012) Are amphitheatre headed canyons indicative of a particular formative process? American Geophysical Union, abstract #EP51A-0969

Turner BR (1983) Braidplain deposition of the Upper Triassic Molteno Formation in the main Karoo (Gondwana) Basin, South Africa. Sedimentology 30:77–89

Twidale CR, Vidal Romani JR (2005) Landforms and geology of Granite Terrains. Balkema, Leiden. 351 pp

Viles HA, Naylor LA, Carter NEA (2008) Biogeomorphological disturbance regimes: progress in linking ecological and geomorphological systems. Earth Surf Proc Land 33:1419–1435

Wessels D, Wessels L (1991) Erosion of biogenically weathered Clarens sandstone by lichenophageous bagworm larvae (Lepidoptera: Psychidae). Lichenologist 23:283–291

3 The Mpumalanga/Limpopo Escarpment: Geology and Fluvial Landforms

Morris Viljoen

Abstract

The Mpumalanga/Limpopo sector of the Great Escarpment of northeast South Africa is unique in its wide variety of rock types which have largely controlled the formation and morphology of the spectacular landscapes of the region. This chapter describes how epeirogenic uplift of the escarpment, followed by headward erosion by rivers into different geological formations, has sculpted different landforms and largely controlled the development of features such as waterfalls, scarp faces, gorges and canyons. Pothole formation in harder rocks has been a major factor in the development of gorges. The weathering characteristics of dolomite formations and nature and origin of dolomitic caves, their dripstone deposits and calcareous tufa deposits along the escarpment region are also described. The spectacular natural features of the region have made it a popular tourist destination.

Keywords

Great Escarpment • Weathering • Erosion • Rivers • Karst

3.1 Introduction

The geomorphic features of the Mpumalanga/Limpopo Escarpment region (also termed the Eastern Escarpment), and indeed of the entire escarpment encompassing southern Africa, are the result of a complex combination of factors. These include isostatic readjustments (uplift and deformation) of the continental crust, largely due to a mantle plume below the African continent; and global climate changes during the Neogene and Pleistocene which resulted in glacioeustatic oscillations in sea level, river incision, and land surface planation (e.g. du Toit 1954; King 1978; Partridge and Maud 2000). These processes have formed macroscale continental erosion surfaces such as the Highveld and Lowveld regions of southern Africa. Subsequently, the dominant processes leading to further planation are those of backward retreat and incision of drainage lines, and the lateral spread of the newly formed valleys (footslopes) at the expense of these former erosion surfaces (King 1978).

The Eastern Escarpment extends for some 300 km from Badplaas in the south to Haenertsberg in the north, and attains altitudes of over 2,000 m a.s.l. A 200 km sector of the escarpment from the Abel Erasmus pass in the north to Badplaas in the south, is discussed in this chapter (Fig. 3.1). Although not as high as the main portion of the Drakensberg Escarpment in KwaZulu-Natal ($\sim$3,400 m a.s.l.; see Chap. 6), the Eastern Escarpment offers a distinct and varied topography, particularly in contrast to the relatively flat, low-lying terrain of the Lowveld (average 400 m a.s.l.) located to the east.

Several prominent fluvial systems dominate the 210 km-long sector of the escarpment that is described in this chapter, including (from south to north) the Komati, Crocodile, Sabie, Blyde and Olifants rivers, which all rise in the high mountain land to the west of the main escarpment edge. These rivers and the geomorphological features found along their courses are associated with significant geotourism attractions such as the world famous Bourke's Luck potholes, Blyde River Canyon, and many scenic waterfalls

M. Viljoen (✉)
University of the Witwatersrand, Johannesburg, South Africa
e-mail: morris.viljoen@vmic.co.za

S. Grab and J. Knight (eds.), *Landscapes and Landforms of South Africa*,
World Geomorphological Landscapes, DOI 10.1007/978-3-319-03560-4_3,

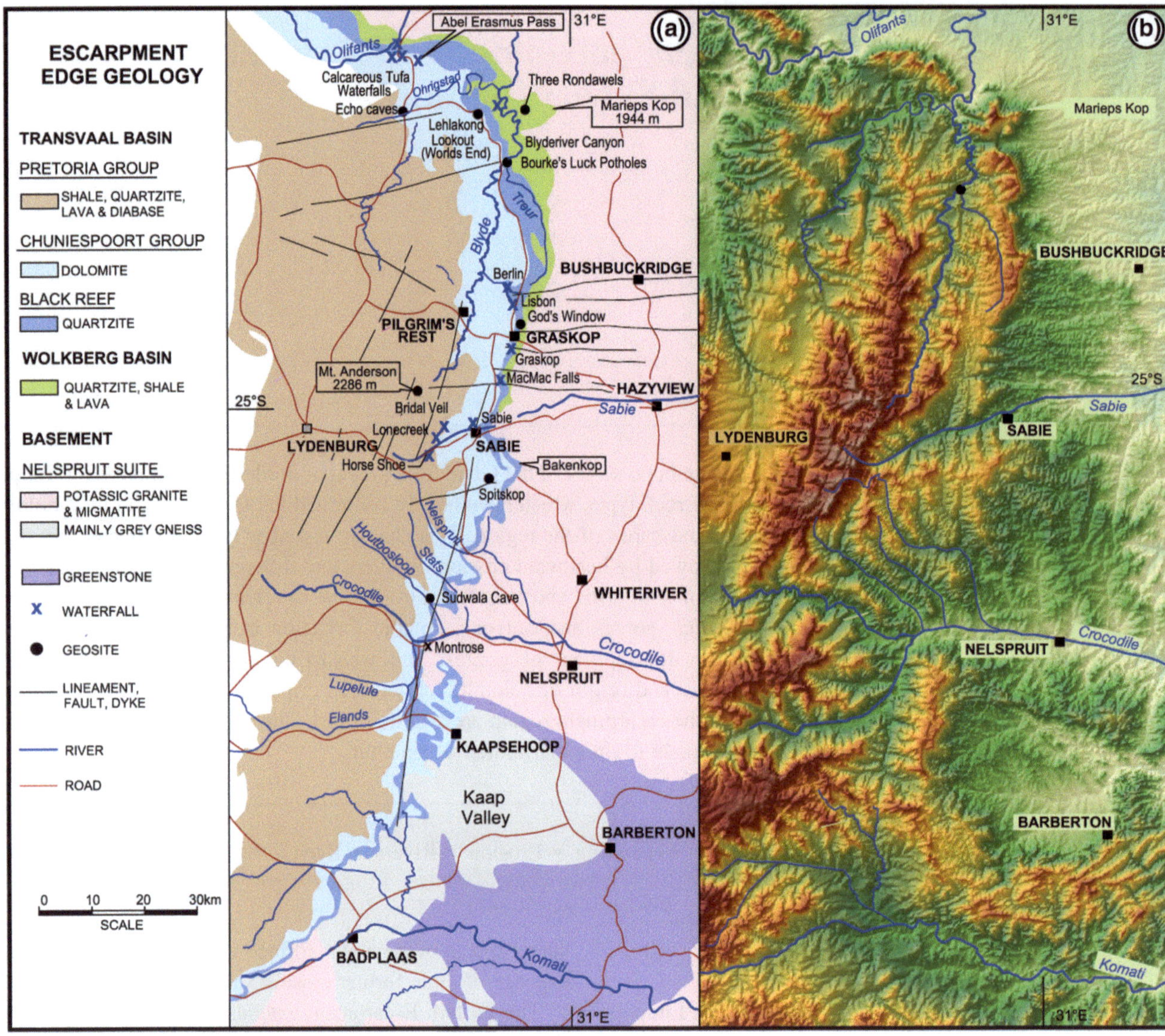

Fig. 3.1 **a** Regional geological and locality map of the central and southern part of the eastern escarpment and Lowveld. **b** Digital elevation model (*DEM*) of the same region, highlighting the relationship between geology and topography

(Fig. 3.1). This chapter provides an account of the dramatic erosional features of the Mpumalanga/Limpopo portion of the Eastern Escarpment (see also Partridge et al. 2010).

3.2 Geological Setting and Evolution

Basement rocks that underlie the Lowveld below the escarpment edge include the ancient volcanic and sedimentary Barberton greenstone belt (which dates to 3,500 million years ago) (Brandl et al. 2006). This greenstone belt forms a conspicuous topographic feature within the Lowveld that almost merges with the escarpment edge in the area west of Barberton (Fig. 3.1). The greenstones were intruded between 3,200 and 3,400 million years ago by granites which were subsequently subject to metamorphism and changed into a range of granitic gneisses. These are less resistant than greenstones and now form the low-lying terrain to the west and south of Barberton and to the east of Mariepskop. A range of more resistant potassium and silica rich granites intruded into the older Basement terrain about 3,100 million years ago and now give rise to the undulating mini-plateau immediately east of the Eastern Escarpment edge between Bushbuckridge and Nelspruit (Robb et al. 2006).

Two younger sedimentary basins overlie the basement rocks to the west, where erosion of the resistant rock formations has formed the dramatic features of the Eastern Escarpment edge, in part controlled by faults and joints (Fig. 3.1). Rocks of the older Wolkberg basin, dated at 2,700 million years ago, are best developed and exposed in the area north of Sabie and forms the dramatic scenery of the Blyde River canyon and Marieps Kop on the escarpment edge, the

Fig. 3.2 Upper surface of small algal stromatolite domes in chert at the entrance to Sudwala caves (*Photograph* M. Viljoen)

latter attaining an altitude of 1,944 m a.s.l. The Wolkberg group comprises a succession of quartzites, shales and lavas.

The resistant Black Reef Formation, dated at 2,650 million years, immediately overlies the Wolkberg sediments and is composed largely of quartzite, which was deposited on an extensive regional unconformity at the base of the overlying Transvaal Sedimentary Basin (Eriksson et al. 2006). The Black Reef Formation largely defines the Eastern Escarpment edge and overlies Wolkberg rocks in the north and basement granitic rocks in the south (south of the town of Sabie) (Fig. 3.1). A thick succession of dolomite with interlayered chert overlies the Black Reef quartzite and is characterised by the presence of stromatolites (fossilised domal algal colonies) which are well exposed south of Sabie and at the Sudwala caves (Fig. 3.2). A resistant chert and breccia layer (the Rooihoogte Formation) marks the top of the dolomite and forms a resistant layer which has influenced subsequent landform development. The higher terrain to the west of the region is overlain by a succession of quartzites and shales of the Pretoria Group.

Fig. 3.3 The spectacular Blyde River Canyon; view from the Lehlakong lookout (World's End) towards 'The Three Rondavels'. Marieps Kop (1,944 m a.s.l.) on the escarpment edge, can be seen in the far distance (*Photograph* M. Viljoen)

3.3 Landscapes and Landforms of the Escarpment Edge

Over the last 170 million years, the escarpment region has undergone ~1,400 m of uplift relative to the Lowveld, as deduced from the elevation of coal deposits at the base of the vast Karoo basin. The coal measures lie at an altitude of 300 m a.s.l. in the Pafuri coalfield of northern Kruger National Park to the northeast, and at 1,700 m a.s.l. on the Highveld plateau north of Middelburg to the southwest. The period of major epeirogenic uplift was followed by land surface denudation with the different rock formations along the escarpment having been etched out and shaped by differential weathering and erosion, in particular by fluvial headward erosional processes.

Fig. 3.4 'Atlas', a dripstone column and archaeological display near Echo cave (*Photograph* M. Viljoen)

3.3.1 Weathering and Erosional Landforms

Resistant quartzite units of the Wolkberg and Black Reef formations form cliff faces (scarp edges) along much of the escarpment edge. Along upper cliff faces, jointed blocks and

Fig. 3.5 Active calcareous tufa deposit with waterfall on Black Reef quartzite; Abel Erasmus Pass (*Photograph* M. Viljoen)

columns of quartzite have detached in many places as a result of preferential weathering and weakening along joints and bedding planes, resulting in these blocks and columns often falling onto lower slopes to form talus accumulations. On the escarpment summit, joints in the quartzites have been preferentially weakened and widened by weathering, with removal of material by river erosion and rockfalls, resulting in distinctive landscapes of isolated pillars, columns and blocks, such as in the Kaapsehoop and Bakenkop areas and along the Treur River (Fig. 3.1). Less resistant shale and volcanic units of the Wolkberg Formation have generally given rise to soil-covered and vegetated slopes, as can be seen where the Blyde River exits the escarpment at the Blydepoort dam. The three 'rondavels' (conical hills) on the east side of the Blyde River Canyon (Fig. 3.3) are good examples of the weathering products of shale, forming the sloping rondavel 'roofs' and more resistant quartzite forming the vertical side 'walls'. A small, horizontal quartzite capping is still present on top of the innermost rondavel and has preserved a conical shale slope below. The two outer rondavels have lost their protective quartzite cappings and the underlying shale 'roofs' form flat domes.

Distinctive features developed on the dolomite terrain include solution furrows and preferentially-weathered dolomite layers between more resistant chert beds. This has also resulted in a distinctive stepped topography on hillslopes. The prominent landmark south of Sabie known as Spitzkop, which was an important locality along the old transport riders' wagon trail between Lydenburg and Delagoa Bay, owes its prominence to a resistant chert capping.

In places, caves have formed by solution of the dolomite along joints, faults and other weak zones. The important Sudwala and Echo caves, both popular tourist attractions, contain impressive dripstone formations and archaeological displays (Fig. 3.4). Calcareous tufa deposits thought to date to about 2 million years ago are widespread in the Blyde Canyon–Abel Erasmus Pass area. These deposits occur where calcium-charged water from the dolomite cascades over knickpoints in the terrain, precipitating calcium carbonate by a combination of oxidation and algal activity. Active tufa deposition can be seen near the Strydom tunnel on the Abel Erasmus Pass (Fig. 3.5) and within the Blyde Canyon itself.

3.3.2 Headward Fluvial Erosional Features: Southern Escarpment

Features formed by headward erosion of rivers are well displayed in the region south of Sabie, where broadly east-west trending tributaries of the Crocodile, Elands and Komati rivers all show how rivers have responded to variations in bedrock properties (Fig. 3.1). The relatively hard Black Reef Formation at the base of the dolomites, together with the chert rich Rooihoogte Formation at the top of the dolomites, has helped arrest river erosion by forming resistant knick points, commonly marked by waterfalls and cascades. The scarp edges have therefore been retreating westwards over time as a consequence of headward river erosion. This is well demonstrated by the Elands River and its tributary the Lupelule, and the Crocodile River and its tributaries the Houtbosloop, Stats and Nelspruit rivers (Fig. 3.1). On the Sabie River and its tributaries, this phenomenon has given rise to the Sabie, Mac Mac and Graskop waterfalls which flow over resistant Black Reef quartzite (Fig. 3.6). These waterfalls have been controlled by the east-west trending Sabie valley dyke swarm and associated fracture zone, which can be clearly identified on the digital elevation model (DEM) of the area (Fig. 3.1). To the west of the town of Sabie, and at the top of the dolomites, a series of waterfalls has formed on the Sabie river and its tributaries. These waterfalls are controlled by the Rooihoogte Formation and include the Bridal Veil, Horseshoe and Lonecreek falls.

Fig. 3.6 The Mac Mac waterfall flowing over a knick point associated with resistant quartzites of the Black Reef and underlying Wolkberg formations. The presence of a number of pronounced east–west trending, vertical joints/fractures which control waterfall and canyon development is evident (*Photograph* M. Viljoen)

Headward erosion has also taken place along north-south trending tributary streams close to the main escarpment edge. These streams, in part, exploit north-south trending lineaments, and resulting erosional patterns are characterised by the almost total isolation of high-standing small plateaus or mesas capped by resistant remnants (outliers) of Black Reef quartzite, sometimes overlain by erosional remnants of dolomite. A good example occurs south of the junction of the Crocodile and Elands rivers where the north-south trending sector of the Elands River, together with its minor east-west tributaries, has isolated a large quartzite/dolomite remnant from the Main Escarpment. The northern face of this plateau, which consists of quartzite overlying granite, is situated a few kilometres southeast of the Montrose waterfall and is clearly visible from the N4 highway. A further example occurs southeast of Sabie where the Bakenkop plateau has been partially isolated by the Spitzkop River, a north-south flowing tributary of the Sabie River. The small quartzite rock pillar known as the 'Pinnacle' at 'God's Window' has been isolated in a similar way by the headward erosion of the Ngwaritsane River and its tributary (Fig. 3.1).

3.3.3 Headward Fluvial Erosional Features: Northern Escarpment

Along the northern sector of the escarpment between Graskop and the Abel Erasmus Pass, only minor east-west trending streams are developed. Instead, the escarpment region is dominated by the Blyde River, a northward flowing tributary of the Olifants River. The Blyde River has its source near the escarpment summit at Mount Anderson (2,286 m a.s.l.), from where it flows across Pretoria Group sediments and then over dolomite before cutting through the underlying Black Reef quartzite at Bourkes Luck to form the start of the spectacular Blyde River Canyon (Figs. 3.1 and 3.3).

The Treur River, a tributary of the Blyde River, has its origin on the escarpment edge north of Graskop and also flows in a northerly direction before joining the Blyde River at Bourke's Luck. At this fluvial junction, the Treur River plunges over rapids at a knick point on Black Reef quartzite, into a mini gorge which widens downstream and within which are found spectacular potholes (Fig. 3.7a, b). Coalescing potholes developed in the quartzite can be

Fig. 3.7 **a** Coalesced potholes cut into bedrock at the base of a mini gorge at the mouth of the Treur River at its junction with the Blyde River. **b** Vertical cylindrical scours with potholes at the base, scoured into the hard, Black Reef quartzite at Bourkes Luck (*Photographs* M. Viljoen)

considered as the final process of gorge formation over long time periods (Fig. 3.7a). Evidence of the vertical boring action of grinding pebbles and boulders within potholes is evident on the side walls of the Blyde River gorge immediately upstream of the Treur River junction (Fig. 3.7b). Today, individual potholes continue to be scoured by the abrasive action of chert and quartzite pebbles trapped within them, particularly during high summer flow and flood events.

3.4 Conclusions

The Mpumalanga/Limpopo Escarpment with its diversity of geology and landforms constitutes a unique sector of South Africa's Great Escarpment. Many of the geological and landscape features are well known tourist destinations with excellent access to viewing points and some existing interpretive centres such as at Bourke's Luck. There are also a number of localities where existing displays could be complemented by more in-depth interpretation of the geology, landscape and scenery for educational and tourism purposes. In this regard, a skywalk is being planned at World's End lookout on the western rim of the Blyde River canyon. The eastern escarpment provides a stark geomorphological contrast to the much lower and flatter Lowveld which includes the Kruger National Park landscape to the east (see Chap. 13), and thus offers visitors one of the world's premier mixed landscape/nature-based tourism hotspots.

Acknowledgments The author would like to thank Profs Stefan Grab, Jasper Knight and Richard Viljoen for critically commenting on and improving this chapter. Lynn Whitfield is thanked for drafting Fig. 3.1.

References

Brandl G, Cloete M, Anhaeusser CR (2006) Archaean greenstone belts. In: Johnson MP, Anhaeusser CR, Thomas RJ (eds) The geology of South Africa. Geological Society of South Africa. pp 9–56

Du Toit AL (1954) The geology of South Africa, 3rd edn. Oliver and Boyd, Edinburgh, 611 pp

Eriksson PG, Altermann W, Hartzer FJ (2006) The transvaal supergroup and its precursors. In: Johnson MP, Anhaeusser CR, Thomas RJ (eds) The geology of South Africa. Geological Society of South Africa. pp 237–260

King L (1978) The geomorphology of central and southern Africa. Monographiae Biologicae 31:1–17

Partridge TC, Maud RR (2000) Macro-scale geomorphic evolution of southern Africa. In: Partridge TC, Maud RR (eds) The cenozoic of southern Africa. Oxford University Press, Oxford, pp 3–18

Partridge TC, Dollar ESJ, Moolman J, Dollar LH (2010) The geomorphic provinces of South Africa, Lesotho and Swaziland: a physiographic subdivision for earth and environmental scientists. Trans Royal Soc S Afr 65:1–47

Robb LJ, Brandl G, Anhaeusser CR, Poujol M (2006) Archaean granitoid intrusions. In: Johnson MP, Anhaeusser CR, Thomas RJ (eds) The geology of South Africa. Geological Society of South Africa. pp 57–94

Landscape and Landforms of the Vredefort Dome: Exposing an Old Wound

4

Roger L. Gibson and W. Uwe Reimold

Abstract

A striking 100 km-long crescent of ridges and valleys straddling the Vaal River along the border between North West and Free State provinces near the towns of Parys and Vredefort is the most obvious remnant of one of the most remarkable geological events in Earth's history. The Vredefort impact event 2,020 Ma ago into the ancient rocks of the Kaapvaal craton is estimated to have left a crater that was originally at least 250 km wide and over 1 km deep. The crater and its infill of broken and melted rocks have long since been stripped away by erosion, rendering the crater margins largely invisible today. However, a central region of rock that was domed upward during the impact event and that bears numerous scars of the catastrophe is still visible. The crescentic Vredefort Mountainland forms a portion of this geological feature, which is referred to as the Vredefort Dome. The landscape of the Dome owes much of its current dramatic topographic relief to the 300 Ma Dwyka glaciation, evidence of which is now being exhumed by the modern Vaal River. Large potholes, sand-blasted rock pavements and the remnants of ancient dune fields testify to more recent shifts in climate in the Mesozoic and Cenozoic. Part of the Vredefort Dome was inscribed as a UNESCO World Heritage Site in 2005.

Keywords

Climate change • Glacial landforms • Meteorite impact structure • Vaal River • Vredefort Dome

4.1 Introduction

Whilst much of the interior plateau of South Africa is dominated by horizontal strata of the 300–180 Ma Karoo Supergroup that have given rise to its characteristic mesa topography (see Chaps. 1 and 2), the northern parts of South Africa expose a far older and more complex geology that, in turn, contributes to a more varied landscape. The Witwatersrand region that extends westwards from Johannesburg (Fig. 4.1) and forms the great watershed of southern Africa is underlain by rocks that formed more than 2,000 Ma ago, among which are gently-dipping resistant white quartzites that mark the signature cuesta ridges of this region (McCarthy and Rubidge 2005). Sandwiched between the Witwatersrand region and the northern edge of the Karoo Basin, however, is a region in which the rocks display complex, locally annular patterns (Fig. 4.1) that, in turn, have produced a highly unusual landscape. The contorted, broken, displaced and strongly rotated rock layers owe their configuration to a remarkable cataclysmic event ca. 2,020 Ma ago, when an approximately 10 km-wide asteroid hit the Earth and formed a gigantic impact crater, the deeply eroded remnants of which can now be seen in the northern Free State, southern North West and Gauteng provinces (Gibson and Reimold 2008; Reimold and Gibson 2010).

R.L. Gibson (✉) · W.U. Reimold
University of the Witwatersrand, Johannesburg, South Africa
e-mail: roger.gibson@wits.ac.za

W.U. Reimold
Museum Für Naturkunde, Berlin, Germany

W.U. Reimold
Humboldt-Universität Berlin, Berlin, Germany

S. Grab and J. Knight (eds.), *Landscapes and Landforms of South Africa*,
World Geomorphological Landscapes, DOI 10.1007/978-3-319-03560-4_4,

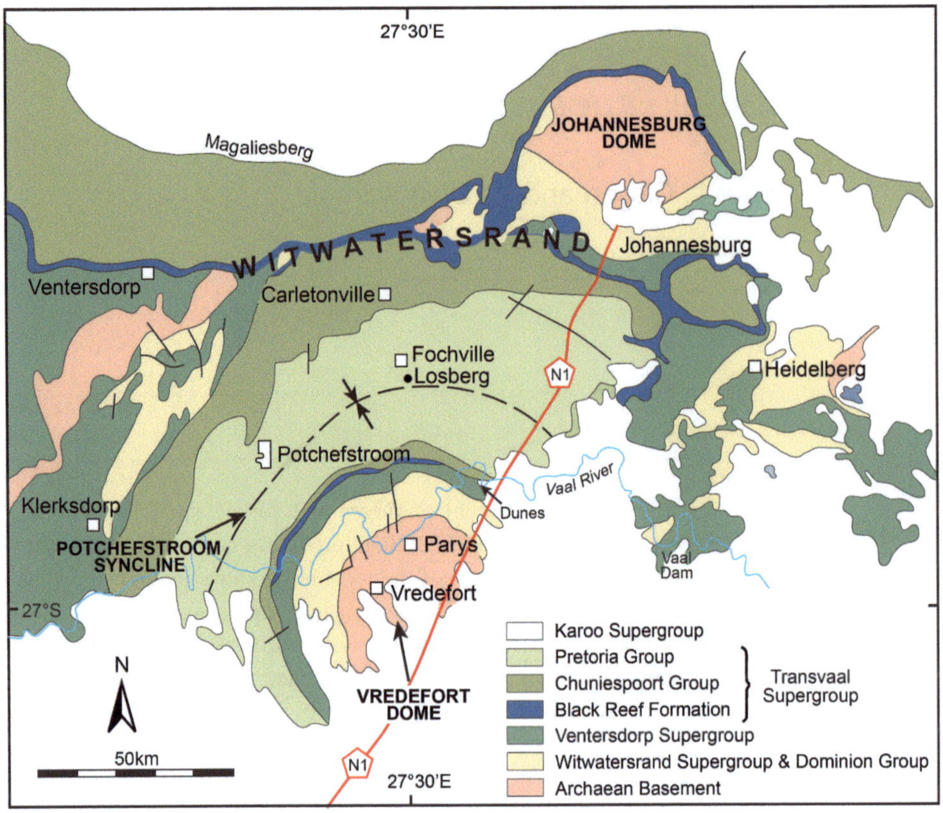

Fig. 4.1 Simplified geological map of the exposed portion of the Vredefort impact structure, with the Vredefort Dome at its centre. The Potchefstroom Syncline marks the outer limit of the Dome. The limits of the original crater may have lain close to, or even beyond, the Witwatersrand. The southern and southeastern parts of the impact structure are hidden beneath a thin veneer of Karoo Supergroup rocks

4.2 Geological Setting

The Vredefort Dome is a region of strongly upturned rock formations centered approximately 15 km south of Parys (Fig. 4.1; Gibson and Reimold 2008). The term "dome" is used here in a geological, rather than geomorphological, sense, in that the landscape is not marked by a central topographic high point; instead, the term describes the arrangement of rock formations that illustrate by both their geometry and their relative ages that old, originally deeply buried, rocks have risen in the centre, pushing aside and rotating the overlying layers (Figs. 4.1 and 4.2). The steep dips of strata within the dome region gradually give way outwards to more gentle dips which are initially centrifugal but, at greater radial distances, switch to centripetal near the towns of Potchefstroom and Fochville (Fig. 4.1). This reversal marks the axial hinge line of the Potchefstroom Syncline, which is the outer limit of the Vredefort Dome. The Dome is thus almost 90 km wide, although most of its southern half presently lies hidden beneath a thin veneer of younger Karoo Supergroup rocks (Fig. 4.1).

The rocks forming the Vredefort Dome span almost one-third of Earth's history (Gibson and Reimold 2008). In the centre lie predominantly granitic gneisses, most of which crystallised from magmas deep below a volcanic arc about 3,100 Ma ago, but which also include even older sea-floor basalts and mudrocks that are now highly metamorphosed gneisses. These coarse crystalline Archaean gneisses define the 40 km-wide *core* of the Dome. Surrounding this core is a 20–25 km wide *collar* of metamorphosed layered sedimentary and volcanic rocks that range in age from 3,074 to 2,100 Ma and that comprise (in order of decreasing age and increasing distance from the centre of the Dome) the Dominion Group and the Witwatersrand, Ventersdorp and Transvaal Supergroups (Gibson and Reimold 2008). The combined original vertical thickness of these rocks is estimated to have been 12–15 km. The Dominion Group and Ventersdorp Supergroup comprise metamorphosed basalt lava flows, whereas the Witwatersrand Supergroup (quartzite, metamorphosed shale, iron formation and conglomerate) and Transvaal Supergroup (dolomite, quartzite and shale) represent sedimentary sequences, but also contain a few thin lava layers, numerous intrusive dolerite sills and several

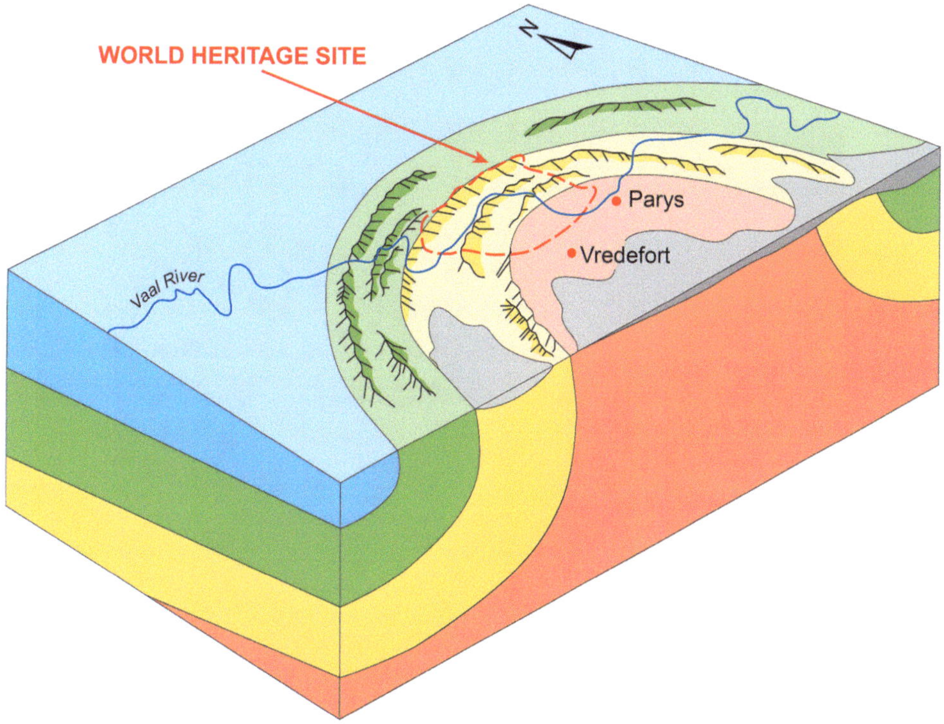

Fig. 4.2 Simplified schematic block diagram of the Vredefort Dome (perspective looking northeast) showing the arrangement of the major rock formations (*orange* = Archaean Basement Complex; *yellow* = Witwatersrand Supergroup and Dominion Group; *green* = Ventersdorp Supergroup; *blue* = Transvaal Supergroup; *grey* = Karoo Supergroup) relative to the Vaal and Vredefort Mountainland. The World Heritage Site occupies only a small portion of the Dome

small granite intrusions. Most of the sedimentary layers have near-vertical dips and are actually inverted in large sections of the collar, so that they dip steeply inwards towards the centre of the Dome (Fig. 4.3a). This overturning reflects the last stage of dome formation, when the dome collapsed downwards and outwards under its own weight. The doming itself was triggered by release of the intense downward compressional forces exerted on the target crust beneath the point of impact. The Dome would have taken no more than a few minutes to form after the impact, but rocks in its core were uplifted by as much as 20–25 km, making its formation one of the most remarkable geological events in Earth's history. It is also one of the few places on Earth where an originally vertical section of Earth's crust more than 20 km deep is now exposed at the surface.

In addition to the obvious structural disturbance of the rock layers (Fig. 4.3a), the rocks in the Vredefort Dome bear witness to the immense forces caused by the impact and its aftermath—distinctive shatter cones and intense fracturing of the rocks, as well as veins and dykes of impact-formed melt rock that are laden with fragments (xenoliths) of the target rocks (Fig. 4.3b), are abundant. Much of the evidence of impact is, however, microscopic-thin (less than 0.001 mm across), closely-spaced, planar lamellae in quartz and zircon crystals, and dense mineral variants (polymorphs) of quartz called coesite and stishovite testify to intense shock pressures that can only be explained by a sudden, violent, high-pressure, shock event. Although an impact origin for the Dome was proposed as early as the 1930s, alternative hypotheses, such as upwelling of a magma diapir or intersecting fault and fold structures, held sway until the latter parts of the last century, particularly among local geologists (see review in Reimold and Gibson 2010).

The 2,020 ± 5 Ma age of the impact event is a composite of the results of several studies based on analysis of U–Pb (uranium and lead) isotope ratios in zircon crystals extracted from a variety of melt-rocks that formed as a result of the impact (Gibson and Reimold 2008).

Another consequence of the impact was intense heating of the rocks by the energy from the shock wave, which metamorphosed and annealed the rocks. The unusual intensity and broad extent of these metamorphic changes (Gibson et al. 1998), together with considerations of the uplift and erosional history of the region (McCarthy et al. 1990), suggest that the rocks presently exposed at the surface were still buried as much as 8–10 km below the surface of the crater immediately after the impact event. Consequently, the modern landscape of the Dome bears no relation to the original crater. The extreme amount of erosion is one of the primary reasons why the decades-long debate about the

Fig. 4.3 **a** View to the east along the inner collar of the Dome, showing the upturned and strongly disrupted quartzite layers of the lower Witwatersrand Supergroup. Dips are generally steep, and directed towards the centre of the Dome (*right*). **b** Large pseudotachylitic breccia dyke in granite from Leeukop Quarry. The *dark grey* matrix is recrystallised melt. It contains fragments of varying sizes derived from the adjacent wallrocks (*Photographs* R Gibson)

original magnitude of the Vredefort impact event and diameter of the crater is unlikely to ever be fully resolved (Gibson and Reimold 2008).

The region contains little evidence of geological activity for almost 1,700 Ma following the impact, until the formation of the Karoo basin. The exceptions are narrow dykes of intrusive granite in the northern collar related to the 1,250 Ma Pilanesberg volcanic complex (see Chap. 5) and a large 1,100 Ma (Reimold et al. 2000) gabbro sill, up to 100 m thick, that is exposed along the Vaal River upstream of Parys. It is likely that the exhumation of the Dome was sporadic in the 2,000 Ma since the crater formed and that periods of erosion were punctuated by periods of burial beneath younger, now vanished, sedimentary sequences.

4.3 Landscape of the Vredefort Dome

The Vredefort Dome is only partially exposed: its southern half is largely covered by a thin veneer of Karoo Supergroup sediments, minor volcanic layers and dolerite, for the most part no more than a few tens of metres thick. What many people mistake for the edge of the Dome is, in fact, the outer edge of the Vredefort Mountainland, which lies between 20 and 30 km from the centre of the Dome, whereas the geological limit—defined as the distance at which the pre-impact sedimentary and volcanic rock layers show no clear structural disturbance or rotation to steeper orientations that can be attributed to the impact—lies 40–45 km from the centre (Potchefstroom Syncline in Fig. 4.1), corresponding to the approximate trace of the Mooi River (Figs. 4.1 and 4.4). Comparison with the well-preserved craters on the Moon and other planets suggests that the original Vredefort crater would have been between 2 and 4 times this diameter, i.e., between 180 and 360 km wide. A 250–300 km diameter is typically favoured, although no direct evidence of the crater margins remains today.

Processed SRTM (Shuttle Radar Tomography Mission) images (e.g. Fig. 4.4) illustrate the contrasting landscapes in the Dome. The core of the Dome is dominated by gently rolling topography at an elevation of approximately 1,410–1,430 m a.s.l., but rising to 1,540 m a.s.l. northeastwards and westwards towards the collar. Topographic relief is more subdued away from the Vaal River towards the south. A few small dome-shaped hills exposing the granite bedrock rise up to 30 m above their surroundings. The Inlandsee Pan, a shallow ephemeral lake, is a well-known local feature formed in a topographic depression 1.5 × 2.5 km wide, at 1,406 m a.s.l.

The Vredefort Mountainland (Afrikaans: *Bergland*), *sensu stricto*, corresponds to the inner and middle parts of the geological collar that lie between 20 and 30 km from the centre of the Dome. The highest point in the inner collar lies at Thwane (1,658 m a.s.l.) in the west-northwestern collar, where the most rugged topography (~150 m vertical relief) is also found; however, the highest point overall (1,676 m a.s.l.) lies north of the Vaal River in the northern collar. Towards the southwest and southeast both the maximum elevation and relief diminish, before the rocks disappear beneath the Karoo rocks.

In contrast to the Witwatersrand region, the ridges of the Mountainland are highly disjointed and, in places, sinuous as a conseqence of fault displacements and buckling caused by stresses during Dome formation (Fig. 4.3a). All ridges

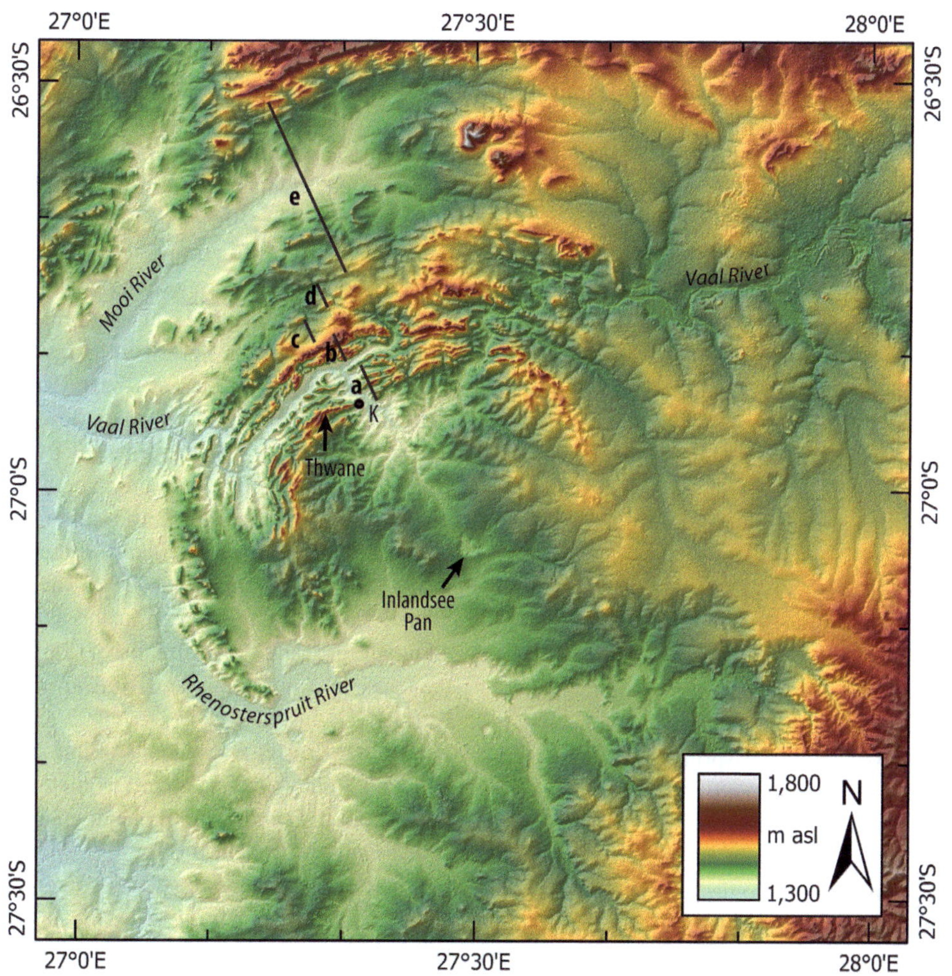

Fig. 4.4 Digital elevation map of the Vredefort Dome from SRTM data (Courtesy M. Rowberry). The image shows the disruption of the normal dendritic drainage pattern of the Vaal River (in the northeast) and Rhenosterspruit River (in the south) by the Dome. The 5 topographic zones (*a–e*) in the collar of the dome and surrounding Potchefstroom Syncline are shown (*K* = Kommandonek)

Fig. 4.5 View towards the northeast along the Venterskroon valley showing a *U-shaped* profile reminiscent of glacial erosion. A small portion of the Vaal River floodplain is seen at *centre-right* (*Photograph* R Gibson)

correspond to outcrops of quartzite, whereas the valleys correspond to meta-shale, meta-basalt and/or meta-dolerite rocks that are more susceptible to chemical weathering. The relative proportions of these rocks play a major role in the type of topography in different parts of the Mountainland, with 5 concentric zones being identifiable beyond the core (Figs. 4.1, 4.2 and 4.4):

(a) the inner collar (lower Witwatersrand Supergroup and Dominion Group), which comprises mostly meta-shale, ironstone, basalt and dolerite sills, with numerous subsidiary thin quartzite bands up to 50 m thick, is marked by sharply-defined ridges of white quartzite and narrow, steep-sided valleys, with the steepest slopes (in places >45°) directed inward;
(b) the much more homogeneous upper Witwatersrand Supergroup, which is predominantly quartzite with a single meta-shale layer, gives rise to a 3–4 km wide zone of extremely rugged topography cut by narrow gorges that exploit faults or major fractures that cut across the strike of the rocks;
(c) the meta-basalt lavas of the Ventersdorp Supergroup have a much more subdued topography, with their weathering resulting in deep, fertile, soils and scattered, broad, rounded hills, rather than ridges;
(d) the dolomite of the lower Transvaal Supergroup (Chuniespoort Group) is poorly exposed, except where it has been impregnated with more resistant chert that gives rise to rubbly linear outcrops; this terrain is characterised by relatively flat topography known for its sinkholes, caves and water springs, which gives the region its name—*Gatsrand*; and
(e) the upper Transvaal Supergroup (Pretoria Group), which straddles the outer limits of the Dome and contains a similar mix of lithologies to that of the lower Witwatersrand Supergroup. Large-scale basinal folding (seen in the curved and sinuous ridges in Fig. 4.4) obscures the concentric arrangement of the layers around the Dome.

4.4 Fluvial Landforms

The Vredefort Dome preserves highly diverse fluvial patterns (Fig. 4.4) that are dominated, but not exclusively controlled, by the Vaal River. The Vaal River has an average gradient that drops ~1 m every 1.5 km across the Dome, but is characterised by a series of local base levels marked by small rapids formed in resistant rocks.

At the macroscale (Fig. 4.1), the Vaal River has a strongly meandering character, despite being incised between 10 and 30 m below its banks. This is consistent with an inherited drainage pattern and explains why the river flows first into and then back out of the Vredefort Mountainland. Despite this, its present path across the Dome is strongly controlled by underlying geology, in this case radial faults related to the impact that cut and displace the collar strata in the northeast, northwest and west, and a concentric foliation in the core gneisses near Parys. Downstream of Kommandonek its northwestward flow across the strata of the inner collar is sharply deflected by the buttress of the Upper Witwatersrand Supergroup and it turns southwestward, parallel to the concentric strike of the collar strata, before exploiting a second fault on the western side of the Dome (Fig. 4.4). The influence of the foliation and fault-related fractures can be seen in the anabranching phenomenon around and downstream of Parys, where the river branches into numerous small linear channels cut into the bedrock.

Tributary rivers and streams in the Dome are ephemeral, although many retain small bodies of standing water throughout the year. Overall, the drainage pattern in the *core* of the Dome is dendritic (Fig. 4.4). Many of the streams in the core of the Dome show strongly sinuous meander characteristics in narrow (usually <100 m wide, but up to 250 m wide in places) marshy ('wetland') overbank areas into which renewed gully incision by up to 2 m has occurred.

The *inner collar* (zones a and b in Fig. 4.4) is dominated by ephemeral streams flowing parallel to the strike of the geological layers and with lower sinuosity than those in the core of the Dome. They are typically incised 3–4 m into thick colluvial deposits containing thin boulder and cobble horizons that locally display a duricrust. In the quartzite-dominated upper Witwatersrand Supergroup (zone b in Fig. 4.4), drainage occurs in linear, narrow, steep-sided, V-shaped valleys and gorges up to several tens of metres deep that are oriented at high angles to the layering and that exploit impact-induced fractures and faults.

Beyond *the Mountainland* at a radial distance of >30 km from the centre of the Dome (zones c, d and e in Fig. 4.4), streams flow either directly into the Vaal (east of the Dome) or into the Rhenosterspruit (south and west) or Mooi River (north), with the latter's course displaying significant geological control.

Between Parys and Kommandonek where it re-enters the collar, the Vaal River exhibits a 7.5 km northwest-flowing reach characterised by significant anabranching, induced by fault-related fractures in the bedrock gneiss and metasedimentary rocks. Midway along this reach, the northeast bank of the river comprises a granite pavement containing large potholes 2–3 m in diameter and over 1.5 m deep. The pavement also displays polish and striations consistent with sand abrasion related to strong northwesterly palaeowinds. The latter may be related to arid climatic conditions that also deposited a small dune field alongside the Vaal River in the northeastern part of the Dome (Fig. 4.1).

4.5 Landforms Likely Associated with Glaciation

In the Dome, the smooth, bare and slightly elongated dome-shaped granite hills of the core locally preserve striations, interpreted as gouges caused by rock fragments being carried in moving ice during the Permian Dwyka glaciations and are, thus, interpreted as glacial features. A large boulder of Witwatersrand Supergroup quartzite several metres across was reported by R.J. Hart (personal communication, 1978) during drilling near the Inlandsee Pan, at least 15 km from the nearest quartzite outcrop, pointing to ice as the likely transport mechanism. The pan itself may be a scooped-out glacial erosion feature. The domical granite hills display limited evidence of exfoliation joints, but the preservation of glacial striae on some surfaces suggests that this is not a particularly intense phenomenon. The Venterskroon valley currently being exploited by the Vaal River has a U-shaped profile consistent with glacial erosion (Fig. 4.5).

4.6 Landscape and Human Dynamics

The Vredefort Dome area contains evidence of human settlement extending back at least several thousand years (e.g. stone tools, rock engravings; Reimold and Gibson 2010) and includes traces of significant settlements from the Late Iron Age up to the early nineteenth century. Human impacts on the regional landscape include gold mining in the collar rocks during the late nineteenth and early twentieth centuries, alluvial diamond diggings along the Vaal River, and quarrying for granite dimension stone in the latter part of the 20th century. Whilst not strictly linked to the Dome itself, South Africa's largest bentonite clay deposits are mined from Karoo rocks overlying the Dome in the south.

In the modern era, a renewed threat of gold mining, using open-cast methods, in the mid-1990s, led local landowners to petition Government for Conservancy status that, ultimately, led to the proposal to declare a portion of the Dome a World Heritage Site. Subsequently, in 2005, a 30,000 ha portion of the Vredefort Dome, extending northwestwards from Vredefort and Parys (Fig. 4.2) was inscribed as a UNESCO World Heritage Site, principally in acknowledgement of its global geological significance as the location of the world's largest and oldest confirmed meteorite impact.

Today, the only continuing mining activity in the Dome area is for sand and bentonite. Large areas are now devoted to private game farms that have restocked the area's wildlife, and to eco- and adventure tourism. Judicious use of the region's resources, particularly in the light of the influx of tourists, is the subject of much discussion, and is complicated by jurisdictional issues between two provinces, four municipalities, and the World Heritage Site Management Authority. Of particular concern is the water quality of the Vaal River, which is affected by activities within the Dome area but most notably from the mining and industrial activities upstream. The World Heritage status has assisted with some projects, such as the clearing of alien vegetation from parts of the Vaal River and upgrading of road infrastructure.

4.7 Summary

The geomorphological evolution of the Vredefort Dome provides snapshots of environmental changes in southern Africa extending back 300 Ma. The present erosion cycle is exhuming an older (Permian) glacial landscape that appears to have been little modified by Mesozoic and Cenozoic erosion related to the Vaal River (Fig. 4.5). Pothole evidence suggests that the Vaal River was once considerably larger than at present; however, sand dunes and sand-blasted rock pavements also indicate that considerably drier conditions prevailed even more recently. Whilst the latter is consistent with increased aridity during cold phases of the Quaternary, it is not clear if the former indicates the waning stages of a humid, tropical climate cycle that dominated in southern Africa for much of the Cretaceous Period, or whether it could be Cenozoic in age. Renewed incision of all rivers in the Dome may reflect the strong Cenozoic uplift within the last 20 Ma, particularly since 5 Ma; however, local river base level migration is also likely to be a contributing factor on a more local scale.

References

Gibson RL, Reimold WU (2008) Geology of the Vredefort impact structure—a guide to sites of interest. Memoir 97 Council for Geoscience, Pretoria, 181 pp

Gibson RL, Reimold WU, Stevens G (1998) Thermal-metamorphic signature of an impact event in the Vredefort Dome, South Africa. Geology 26:787–790

McCarthy TS, Rubidge BS (2005) The story of earth and life, a southern African perspective on a 4.6-billion-year journey. Struik Publishers, Cape Town, 333 pp

McCarthy TS, Charlesworth EG, Robb LJ (1990) Geological studies related to the origin of the Witwatersrand Basin and its mineralisation—an introduction and strategy for research and exploration. S Afr J Geol 93:1–4

Reimold WU, Gibson RL (2010) Meteorite impact! the danger from space and South Africa's mega-impact, the Vredefort structure, 3rd edn. Springer, Berlin, 319 pp

Reimold WU, Pybus GQJ, Kruger FJ, Layer PW, Koeberl C (2000) The Anna's Rust Sheet and related gabbroic intrusions in the Vredefort—part of a widespread Kibaran magmatic event in the Kaapvaal Craton and beyond? J Afr Earth Sci 31:499–521

Geomorphological Evolution of the Pilanesberg

5

R. Grant Cawthorn, Jasper Knight, and Terence S. McCarthy

Abstract

The geomorphological evolution of the Pilanesberg Complex, an alkaline igneous body emplaced around 1,250 million years ago and subsequently exhumed by erosion, reflects the interplay of tectonic and climatic factors over very long time periods that have given rise to regional-scale patterns of geomorphological processes and resultant landforms. The topography and river patterns of the Pilanesberg region have evolved by progressive denudation of Karoo sedimentary rocks (<30 Ma), revealing evidence for a much older, relict land surface that may have been formed during the Dwyka glaciation around 300 million years ago. The absence of significant placer deposits of gold, platinum and chromium is explained by this proposed model.

Keywords

Denudation • River patterns • Alluvial fans • Igneous intrusion • Dwyka glaciation • Placer deposits

5.1 Introduction

Over long time scales and large spatial scales, the evolution of the South African landscape reflects the interplay between tectonic factors, including magmatism and epeirogenic processes that contribute to land surface uplift and creation of relief, and land surface processes of denudation and sediment transport that respond to this geologic forcing and which are also strongly driven by climate (Partridge and Maud 1987). The Pilanesberg (Fig. 5.1) is a spectacular geomorphological feature forming the heart of the Pilanesberg National Park (established 1984) in which there are close connections between geology, geomorphology, ecology and archaeology, and the Park today supports a wide range of wildlife and ecosystems (Carruthers 2011). This chapter aims to (1) outline the geologic setting and formation of the Pilanesberg Complex, (2) describe patterns of post-emplacement erosion and development of the regional pre-Karoo land surface, and (3) discuss the geomorphology of the Pilanesberg, in particular hillslope evolution, and river and valley-fill patterns.

5.2 Geological Setting and Landscape History

5.2.1 Geologic History

The north-eastern portion of South Africa is underlain by the Kaapvaal Craton which grew by microcontinent accretion between ~3,600 and 3,000 Ma (McCarthy and Rubidge 2005). Since then, the craton has formed a stable platform upon which various sedimentary and volcanic rocks were deposited. The most important of these are the Pongola and Dominion Groups between 3,000 and 2,900 Ma, the Witwatersrand and Ventersdorp Supergroups between 2,900 and 2,700 Ma and the Transvaal Supergroup between 2,600 and 2,100 Ma. The Transvaal Supergroup rocks include the Magaliesberg Quartzite, discussed later. About 2,060 million years ago, the Transvaal rocks were invaded by magmas which formed felsic

R.G. Cawthorn (✉) · J. Knight · T.S. McCarthy
University of the Witwatersrand, Johannesburg, South Africa
e-mail: Grant.Cawthorn@wits.ac.za

S. Grab and J. Knight (eds.), *Landscapes and Landforms of South Africa*,
World Geomorphological Landscapes, DOI 10.1007/978-3-319-03560-4_5,

Fig. 5.1 View looking to the northwest of the mountainous Pilanesberg (horizon) rising from the flat plain of the Bushveld (foreground) (*Photograph* R.G. Cawthorn)

lavas, the Bushveld Complex and granites. Slow crystallization of Bushveld Complex magmas resulted in the differential segregation of minerals, forming the world's largest deposits of chromium, platinum group metals and vanadium. Lithospheric subsidence allowed accommodation space for accumulation of the Waterberg Group that ended 1,750 million years ago. Thereafter, geological activity more or less ceased, except for the Pilanesberg event at about 1,250 Ma, until 300 Ma, when accumulation of the rocks of the Karoo Supergroup commenced.

The deposition of the Karoo Supergroup is particularly relevant to the geomorphology of the Pilanesberg. About 300 million years ago that portion of the Gondwana continent which is today South Africa, lay over the South Pole and was covered by the Dwyka glaciation ice sheet which eroded the land surface. In areas where the rock layers were of similar hardness, the eroded glacial surface was relatively flat, whereas rock layers of contrasting hardness resulted in ridges and valleys. South Africa emerged from beneath the ice sheet as Gondwana slowly moved north, and its glaciated surface became buried by Karoo Supergroup rocks, including temperate sandstones and mudstones (Ecca and Beaufort Groups) and desert sandstones (Stormberg Group). The eruption of Drakensberg basaltic lavas, which buried the South African portion of Gondwana and heralded continental breakup, abruptly terminated sediment accumulation 180 million years ago. With the break-up of Gondwana, a new cycle of erosion commenced, which removed much of the Karoo Supergroup cover over the northern Kaapvaal Craton, and exposing the pre-Karoo land surface. The processes operative during the last 90 million years have been summarized by McCarthy (2013), and the effects of this erosion and exhumation are seen in the Pilanesberg region today.

5.2.2 Landscape and Park History

The unusual rock types of the Pilanesberg Complex have resulted in distinctive patterns of soils and drainage, described in Sect. 5.4. Human occupation over long time

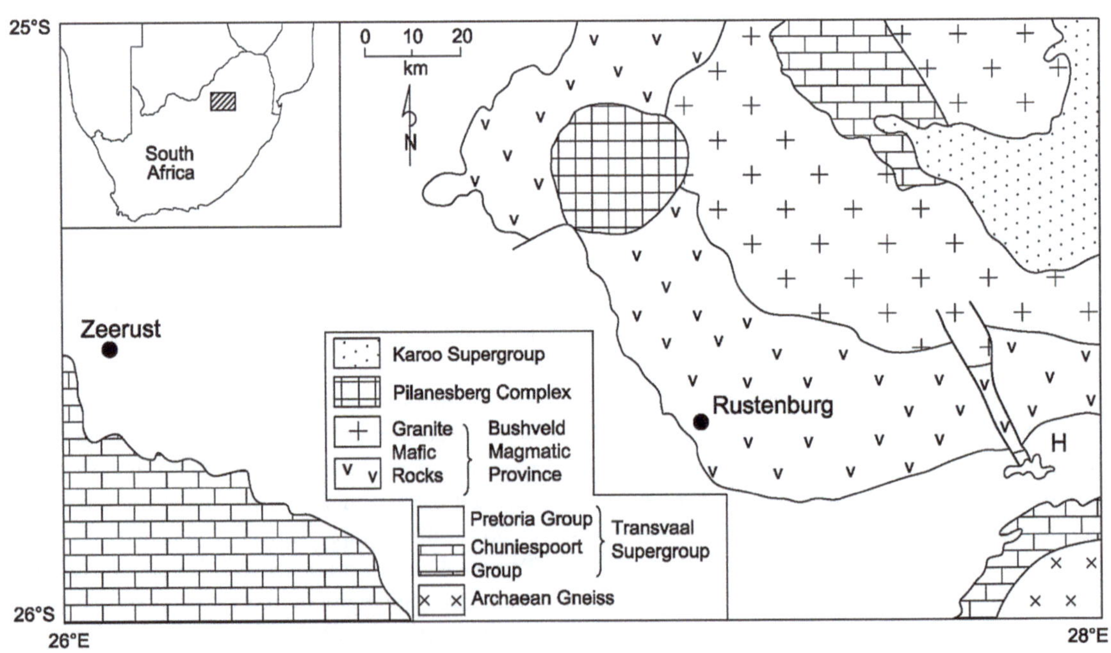

Fig. 5.2 Simplified geological map of the area surrounding the Pilanesberg Complex. *H* = Hartbeespoort Dam

Fig. 5.3 Oblique aerial view of the south-west portion (looking north-west) of the concentrically arranged layers of rocks of the Pilanesberg Complex (*Photograph* R.G. Cawthorn)

periods has exploited these natural resources. Middle Stone Age artefacts are recorded throughout the Pilanesberg. Pastoralism and metal production during the later Iron Age was associated with development of larger, settled Batswana chiefdom communities. Later, Setswana-speaking groups, who built cattle kraals across the region, were succeeded by the invading Khamalo tribe which was succeeded in turn by white Voortrekker settlers in the 1840s who established the Pilanesberg as a farming area. In the 1970s, an agreement between white farmers and the regional government of the then Bophuthatswana established the region as an area of conservation and tourism. The Pilanesberg conservation area was established in 1979 and formalised as a National Park in 1984. Farm buildings and infrastructure were removed, and game was reintroduced to the area (see Carruthers 2011). The general absence of surface water across the Pilanesberg was an important limitation on agriculture: only one significant dam (Mankwe) was built by early settlers, and for much of the year, surface water was scarce. Since becoming a National Park, newer dams have been built for species such as Hippo, but this can also result in an ecological imbalance against drought-tolerant species.

The Pilanesberg rocks are alkaline and groundwater is consequently saline. Such water is palatable to natural game, but less so for domesticated animals and humans, rendering farming extremely challenging in the early twentieth century. There are a number of small fluorite occurrences in the

Fig. 5.4 Photographs of landscape geomorphology in the Pilanesberg. **a** Outcrop of red syenite providing a slope sediment source. **b** Incised V-shaped valley terminating in a colluvial fan. **c** Mixed bedrock/sediment lowland river. **d** Dry ephemeral river bed (*Photographs* J. Knight)

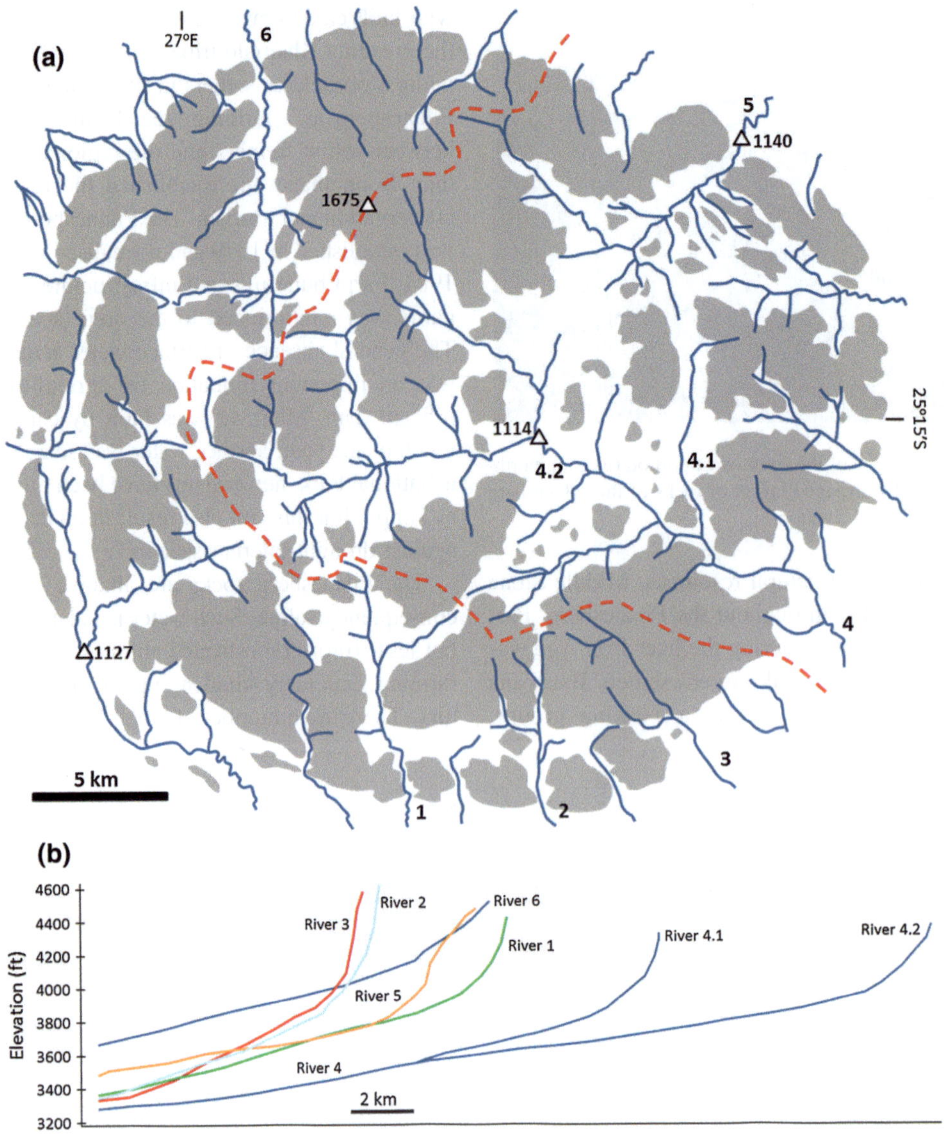

Fig. 5.5 **a** Simplified distribution of hills (*shaded*) and valleys (*not shaded*) in the Pilanesberg showing the outward flowing rivers and principal watershed (*dotted line*). Spot heights (*triangles*) are given in metres. **b** Long profiles of selected rivers (marked 1–6) within the area marked in (**a**). Note the stepped profiles of some rivers (e.g. river 5) and that the low-relief landscape lying outside of the Pilanesberg broadly slopes from north to south, opposed to regional river flow as shown in Fig. 5.6

park, one of which was exploited and the old pit and waste dumps remain visible from park roads. Groundwater concentration of fluorine is well beyond health limits; consequently, local inhabitants often show dental discoloration.

5.3 The Pilanesberg Event

The Pilanesberg Alkaline Province represents both intrusive and extrusive igneous activity that took place in the north-western Kaapvaal craton and areas eastwards around 1,250 million years ago (Verwoerd 2006). The Pilanesberg Complex itself is 25 km in diameter (Fig. 5.2) and is one of the largest such alkaline bodies in the world. Intrusion of these igneous rocks was probably aided by regional tensional forces that also produced north-northwest orientated dykes of syenite and alkali gabbro that extended for about 400 km from the Witwatersrand to Botswana. Shand (1928) produced a detailed geologic map of the Pilanesberg Complex that has been largely confirmed by subsequent studies. Shand (1928) suggested it was a shallow lopolith, whereas Lurie (1973) suggested it was a tilted series of cone sheets which were intruded into a volcanic capping.

The volcanic rocks consist of very fine-grained trachytes and phonolites with sanidine phenocrysts as the only visible phase. The intrusive rocks are chemically comparable to the

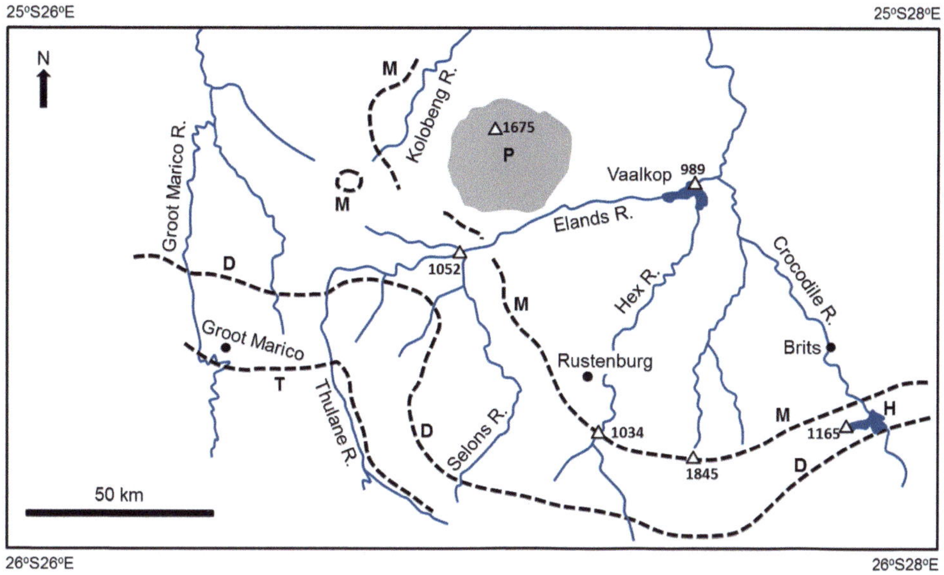

Fig. 5.6 Map of the main north-flowing rivers around the Pilanesberg (*P*). Spot heights (*triangles*) are given in metres. Ridges (*dashed lines*) developed in quartzites of the Transvaal Supergroup are *T* Timeball Hill, *D* Daspoort and *M* Magaliesberg. Between the ridges lies low ground underlain by shales. *H* Hartbeespoort Dam. (Compare with the map of Wellington (1937), his Fig. 5.1, showing the area east of that shown here and covering Pretoria and Johannesburg.)

lavas but are coarser grained and formed variably coloured syenites and foyaites. Cauldron subsidence probably resulted in the downward collapse of the entire structure, preserving a near-perfect circular shape within the surrounding Bushveld Magmatic Province.

5.4 Geomorphology of the Pilanesberg

5.4.1 Topography

The Pilanesberg defines a near-perfect circle, with near-continuous concentric rings of hills (Fig. 5.3), each formed of a different rock type, and with intervening valleys. The entire structure ranges in height from 100 to 500 m above the surrounding flat landscape, but the highest point (~1,560 m) is broadly accordant with that on the Magaliesberg to the southeast. In more detail, the hills of the Pilanesberg individually have a relative relief of 30–200 m and comprise rectilinear slope elements with prominent basal breaks of slope and V-shaped valleys with underfit rivers. Slopes show a discontinuous cover of weathered bedrock material and thin soils, which is typical of weathered alkaline bedrock (Fig. 5.4a). Large residual boulders commonly litter bedrock slopes and valley foot locations and likely reflect colluvial stripping of weathered fragments leaving larger boulders behind as an erosional lag. In contrast, outside of the Pilanesberg, on the Bushveld Complex, soils are thick, dark and formed of clay from weathered gabbros.

5.4.2 Basins and Valley Fills

Within the Pilanesberg itself, semi-enclosed internal basins on the order of 5–100 ha in area are formed where constricted valley mouths and interlocking spurs help capture sediment within small upland catchments, leading to aggrading valley floors separated by steeper river outfalls. Sediments coming into these small internal basins are derived by weathering followed by mass movement and creep from surrounding hillslopes. This has meant that sediment from weathering within the Pilanesberg is largely retained within local depositional basins and is not lost beyond the complex to out-flowing rivers. This inference indicates that most of the valleys in the Pilanesberg are pre-Karoo in age and were originally then filled with Karoo sedimentary rocks.

Hillslopes show linear erosional scours up to 20 m deep that start below the hillcrest and terminate above a prominent basal break of slope where they shallow and grade into small fans (Fig. 5.4b). These fans represent coalescent alluvial and colluvial fans that together form shallowly sloping (at 2–3°) valley fills. Where exposed, the fills are spatially variable in thickness but can attain 4–8 m, are composed of massive to stratified coarse-grained colluvium, and do not represent alluvial floodplain deposits.

5.4.3 Rivers and Drainage Patterns

The Pilanesberg has a summer-dominated rainfall of 500 mm per year and most streams are ephemeral. It is

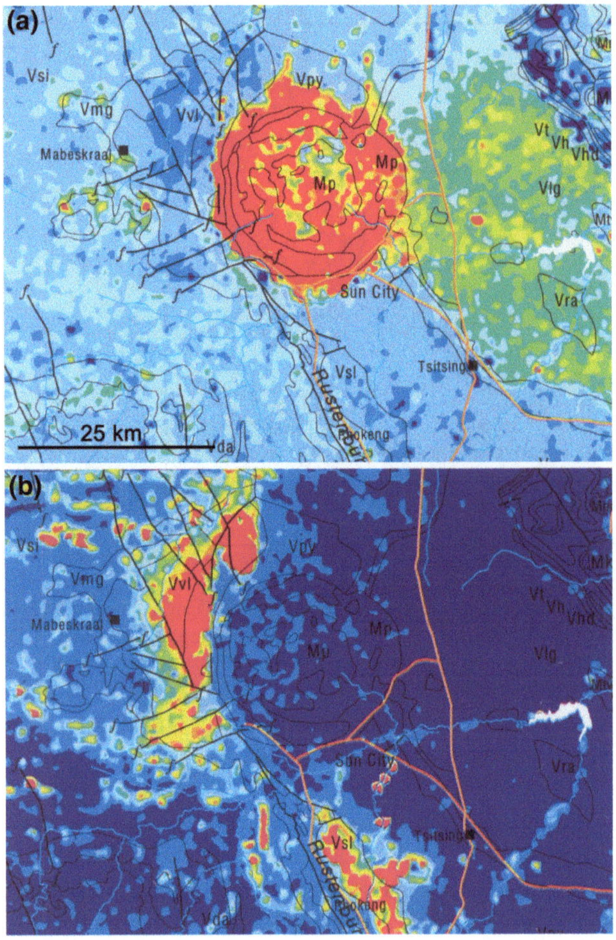

Fig. 5.7 Soil geochemical maps of the Pilanesberg area (from Wilhelm and van Rooyen (2001), with permission from the Council for Geoscience); the scale is the same for both maps. **a** Thorium content values (where *red colour* >40 ppm) are extremely high in the Pilanesberg, but thorium is not being transported by outflowing rivers; **b** Chromium content (where *red colour* >5,000 ppm) is especially abundant in specific layers in the Bushveld Complex surrounding the Pilanesberg, but likewise chromium is not being transported in lithology-crossing rivers. The circular outline of the Pilanesberg Complex is clearly delimited by the high Th values in (**a**)

notable that river drainage patterns are different inside and outside of the Pilanesberg. In part, this is due to topographic differences, but it also strongly reflects geologic history. Groundwater is extremely saline and rich in fluorine and is unsuitable for any form of domestic agriculture.

Within the Pilanesberg, river patterns are eccentric with many doglegs, indicative of geologic control on the position of hills and valleys (Fig. 5.5). Perennial rivers are present only at lower elevations; ephemeral rivers are common on slopes and are likely associated with slope mass movements (Fig. 5.4c, d). Riverbeds are mainly floored by bedrock with sand and gravel patches located within pools. These rivers are only a few metres wide and are not generally associated with wide alluvial floodplains or fine-grained overbank deposits. Throughout, rivers are underfit with laterally restricted and compartmentalised reaches that are separated from each other by valley spurs. Based on their small size and general absence of fluvial sediments, the present rivers are not considered a significant geomorphic agent except during episodic floods when hillslopes are likely reactivated, building aggradational colluvial fans that merge laterally towards the lowland river systems. It is likely that this downslope sediment supply was much higher during previous more humid phases, consistent with evidence elsewhere across southern Africa (e.g. Temme et al. 2008). One such fan containing boulders up to 30 cm in diameter can be seen just outside the Pilanesberg where a major road cuts the outflow of stream number 4 (Fig. 5.5).

Outside the Pilanesberg, rivers generally flow from south to north (Fig. 5.6), but have locally been deflected eastwards by the emergence of the Pilanesberg by erosion. River system evolution around Johannesburg and Pretoria has been described by Wellington (1937, 1941), and we extrapolate his findings to the present area (Fig. 5.6). The main rivers all flow northwards and cut across the topography of alternating quartzites and shales of the Transvaal Supergroup. Wellington concluded that these directions were superimposed as a result of erosion of overlying softer Karoo sedimentary rocks. A notable exception is the Elands River which flows due east before it reaches the Vaalkop dam. It is surmised that the original flow direction of the Thulane and Selons rivers was north and probably continued into the Groot Marico and Kolobeng rivers, respectively, when there was still surviving Karoo sedimentary rock in that area. As the Karoo was eroded, a barrier of hard Magaliesberg Quartzite emerged west of the Pilanesberg (elevation of 1,140 m) and erosion was more rapid to the east (elevation of 1,000 m at Vaalkop Dam). As a result, the east-flowing Elands River developed, which eventually captured these two rivers. Thus, the Elands River defines a course influenced by present patterns of exposed geology rather than being inherited from palaeo-rivers flowing over Karoo rocks.

5.5 Denudation History and the Pre-Karoo Land Surface

Large-scale continental flexure (Burke and Gunnell 2008) may have been the trigger for regional river downcutting into the African Surface at around 30 Ma (Partridge and Maud 1987), which is approximately the level of the present Magaliesberg and Pilanesberg summits. In the Pilanesberg, the period 30–0 Ma may correspond to the erosional removal

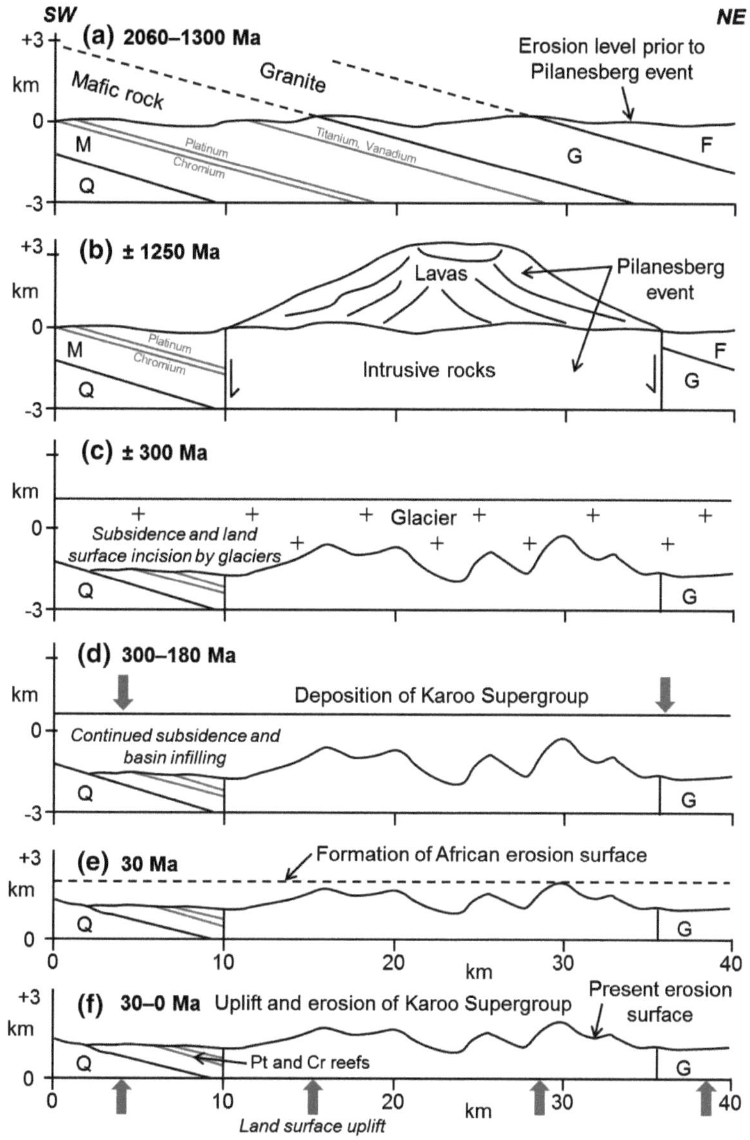

Fig. 5.8 Schematic cross section from SW to NE showing the stages of evolution of the Pilanesberg land surface. Approximate ages and time periods for these sequences are indicated. **a** Conditions prior to the Pilanesberg event; **b** intrusives and extrusives associated with the Pilanesberg event; **c** incision and creation of relief by Dwyka age glaciers; **d** deposition of Karoo Supergroup sediments; **e** erosion of the Karoo Supergroup to the level of the African erosion surface; **f** land surface uplift followed by erosional stripping of Karoo sediments revealing the present land surface. *Q*, *M*, *G* and *F* refer to Magaliesberg Quartzites, and mafic, granitic and felsic rocks of the Bushveld magmatic province, respectively

of Karoo cover rocks (maybe >600 m thickness), exhuming a pre-existing and buried topography, with a superimposed drainage system (Wellington 1937, 1941). It is also relevant to consider the nature of this eroded material. The Pilanesberg Complex is geochemically distinct from surrounding country rocks of the Bushveld Magmatic Province. As a result, any river sediments exported from the Pilanesberg Complex should be chemically identifiable. Systematic regional soil surveying shows that thorium (Fig. 5.7a), niobium and other geochemical species derived from the Pilanesberg are not transported either long distances or in large amounts (Labuschagne et al. 1993; Wilhelm and van Rooyen 2001). It is the Karoo rocks that were eroded during this period. It is thus evident that very little of the Pilanesberg rocks has been eroded during the development of the present-day land surface.

The same lack of river sediment transport is also characteristic of rivers crossing the Bushveld Magmatic Province, which hosts more than half of the global known platinum, chromium and vanadium. Platinum, chromium

(Fig. 5.7b) and ilmenite (from vanadiferous ores) are highly durable, both chemically and mechanically, and so their general absence as placers in streams that straddle source reefs in the Bushveld Complex (Wilhelm and van Rooyen 2001) is truly remarkable. The same applies to the near-complete absence of gold in recent placers on the Witwatersrand, despite draining the world's largest gold deposits.

Here, we extend the mechanism of landscape denudation proposed by Wellington (1937, 1941) to the Pilanesberg (and Witwatersrand) area (Fig. 5.8). The total absence of placer evidence for gold, platinum, chromium and titanium is a crucial observation and suggests that this area was scoured by Dwyka-age glaciers which eroded and transported ore minerals great distances off the present-day land surface of South Africa (Barath and Dunlevey 2010; van Schijndel et al. 2011). We therefore infer that the present topography is essentially a relict glaciated landscape produced by the Dwyka glaciation, which was subsequently covered by Karoo sediments. Erosion of these much softer rocks has revealed the older land surface (Fig. 5.8).

5.6 Summary

The geomorphology of the Pilanesberg reflects the interactions between igneous and climatically driven processes over long time periods and that today's regional landscape is dominantly an erosional relict. It is also notable that present rivers are underfit and valley fills are driven by slope and not fluvial processes. The possibility that the Pilanesberg was shaped mainly by the Dwyka glaciation can help explain the absence of mineral placers in the region. The development of the Pilanesberg as a National Park highlights the interrelationships between geology, geomorphology and ecosystems, which was central to the purpose of the establishment of the National Park in 1984.

Acknowledgments We thank two anonymous reviewers for their comments.

References

Barath S, Dunlevey JN (2010) Gold in Karoo supergroup heavy-mineral deposits, KwaZulu-Natal, South Africa. S Afr J Geol 113:437–444

Burke K, Gunnell Y (2008) The African erosion surface: a continental-scale synthesis of geomorphology, tectonics, and environmental change over the past 180 million years. Geol Soc Am Memoir 201:66 pp

Carruthers J (2011) Pilanesberg National Park, North West Province, South Africa: uniting economic development with ecological design—a history, 1960s to 1984. Koedoe 53(1), Art. #1028, 10 pages. doi:10.4102/koedoe.v53i1.1028

Labuschagne LS, Holdsworth R, Stone TP (1993) Regional stream sediment geochemical survey of South Africa. J Geochem Expl 47:283–296

Lurie J (1973) The Pilanesberg: rare earth element geochemistry and economic potential. Unpublished PhD thesis, Rhodes University, 308 pp

McCarthy TS (2013) The Okavango Delta and its place in the geomorphological evolution of Southern Africa. S Afr J Geol 116:1–54

McCarthy TS, Rubidge B (2005) The story of earth and life. Struik Nature, Cape Town, 334 pp

Partridge TC, Maud RR (1987) Geomorphic evolution of southern Africa since the Mesozoic. S Afr J Geol 90:179–208

Shand SJ (1928) The geology of Pilanesberg (Pilaan's Berg) in the western Transvaal: a study of the alkaline rocks and ring-intrusions. Trans Geol Soc S Afr 31:97–156

Temme AJAM, Baartman JEM, Botha GA, Veldkamp A, Jongmans AG, Wallinga J (2008) Climate controls on late Pleistocene landscape evolution of the Okhombe valley, KwaZulu-Natal, South Africa. Geomorphol 99:280–295

van Schijndel V, Cornell DH, Karlsson L, Olsson JR (2011) Baddeleyite geochronology and geochemistry of mafic cobbles from the Dwyka diamictite: new insights into the sub-Kalahari basement, South Africa. Lithos 126:307–320

Verwoerd WJ (2006) The Pilanesberg alkaline province. In: Johnson MR, Anhaeusser CR, Thomas RJ (eds) The geology of South Africa. Geological Society of South Africa, Johannesburg/Council for Geoscience, Pretoria, pp 381–394

Wellington JH (1937) The pre-Karroo peneplain in the south-central Transvaal. S Afr J Sci 33:231–295

Wellington JH (1941) Stages in the process of river-super-position in the southern Transvaal. S Afr J Sci 37:78–96

Wilhelm HJ, van Rooyen RC (2001) Geochemical map series, 1:1,000,000 scale. 2526 Rustenburg. Council for Geoscience, Pretoria

6 The Drakensberg Escarpment: Mountain Processes at the Edge

Jasper Knight and Stefan Grab

Abstract

The geologic history of the Great Escarpment, which includes within it the Drakensberg escarpment, closely follows cycles of tectonic evolution and land surface denudation from the Jurassic to Miocene that affected the entire southern African region. Along the Drakensberg escarpment, which includes some of the highest mountain summits in southern Africa, the presence of flat-lying Jurassic basalts has strongly influenced processes and patterns of subsequent weathering, erosion and mass movement, in particular during cooler climate periods of the Quaternary. Distinctive weathering, periglacial and glacial, mass movement and fluvial phenomena have resulted from this interplay between geology, climate and geomorphological processes over the Quaternary and Holocene. The Drakensberg escarpment region also shows a close interconnection between landscape geomorphology, biodiversity and patterns of human cultural occupation during the Holocene. For these reasons, conservation of geomorphological, ecological and cultural sites in the Drakensberg escarpment, and their sustainable management under conditions of climate change, is an important contemporary issue.

Keywords

Landforms • Quaternary climate change • Denudation • Drakensberg • Basalt

6.1 Introduction

The Great Escarpment occupies a key landscape position, extending as an arc for several 100 km within the interior of southern Africa and marking an important physiographic boundary, separating coastal lowlands and fold mountains around the perimeter of southern Africa from the high plateau and semi-arid river systems of the interior (Moore et al. 2009; Partridge et al. 2010) (Fig. 6.1). Furthermore, the escarpment represents a topographic, geomorphic, climatic and ecological boundary, and thus its landscapes and evolutionary patterns and processes are in many cases quite distinct from those of surrounding areas. Geomorphologists like Lester King and Walther Penck also used evidence from the Great Escarpment and adjacent areas to devise models of long-term landscape evolution (King 1953). Due to its scale and diverse array of physical environments, most studies have focused on only parts of the Great Escarpment, in particular the Drakensberg escarpment complex, which is the central eastern and highest portion of the Great Escarpment. The Drakensberg area is of particular interest because here the escarpment reaches over 3,400 m a.s.l. (highest peak at 3,482 m a.s.l.) and shows outward-facing near-vertical scarps up to 600 m high (Fig. 6.1b). As a consequence of this prominent relief and high orographic lapse rates, temperature and precipitation regimes vary substantially between footslope and scarp edge locations (Grab 2013). In turn, this has resulted in different geomorphological processes dominating in different topographic contexts (Fig. 6.1b). This chapter focuses on the Drakensberg, so named by the Dutch settlers and meaning 'Dragon mountain', while to the local Zulu these mountains have long been

J. Knight (✉) · S. Grab
University of the Witwatersrand, Johannesburg, South Africa
e-mail: jasper.knight@wits.ac.za

S. Grab and J. Knight (eds.), *Landscapes and Landforms of South Africa*,
World Geomorphological Landscapes, DOI 10.1007/978-3-319-03560-4_6,

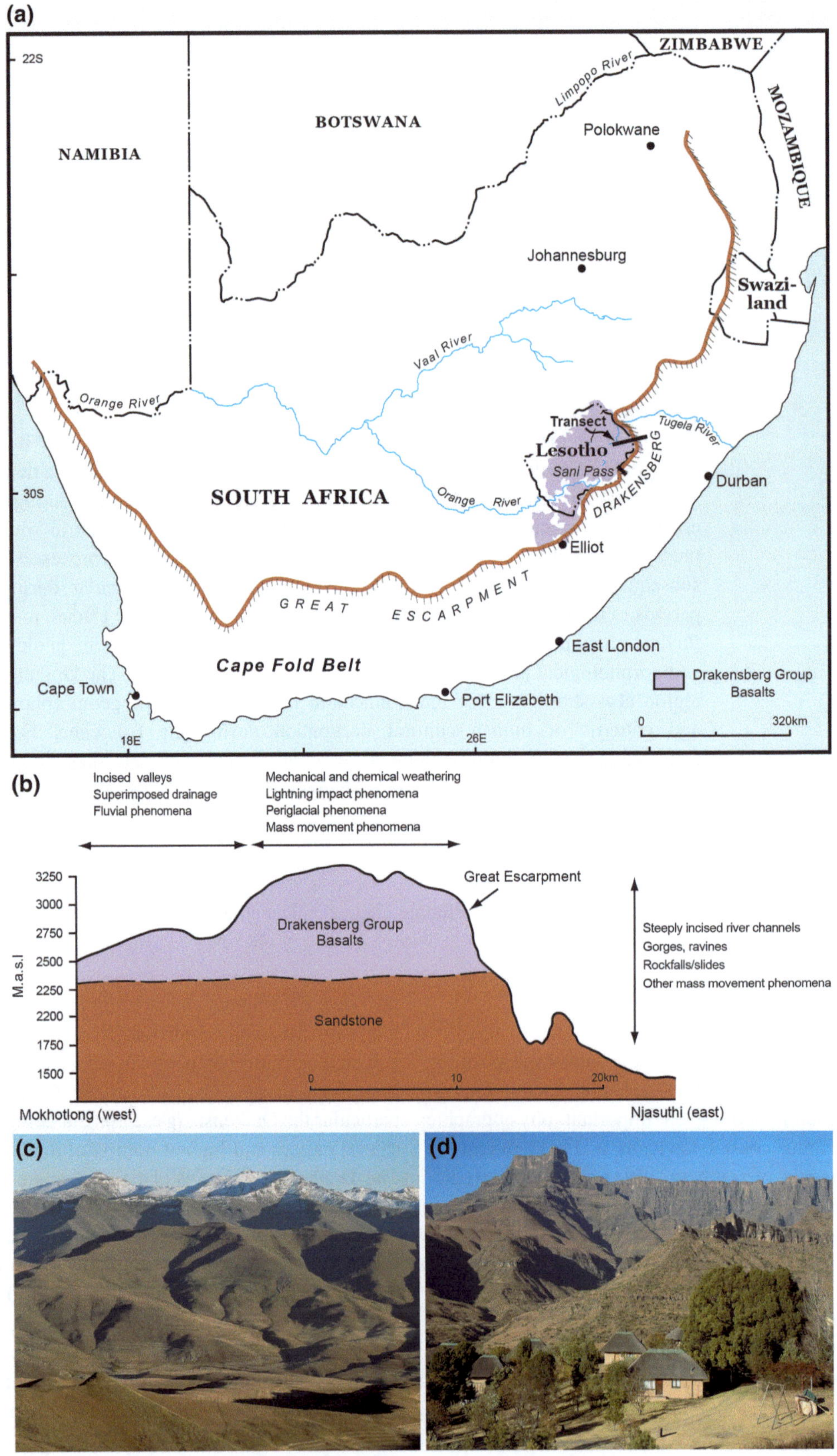

Fig. 6.1 **a** Location of the Great Escarpment across southern Africa and the Drakensberg escarpment sector which is described in detail here; **b** west–east topographic cross section through the Drakensberg escarpment showing the contrast in relief; **c** landscape view west of the escarpment edge in eastern Lesotho; **d** view of the eastern edge of the Drakensberg escarpment (*Photographs* S. Grab)

Fig. 6.2 The Great Drakensberg Escarpment displaying its 'barrier of spears' and vertical cliffs which reach 300–600 m relative relief in places (*Photographs* S. Grab)

known as uKhahlamba (which aptly means 'Barrier of Spears'; Fig. 6.2a, b). Here, we describe the landforms and geomorphic processes typical to the foothills and high summits. Precipitation is highly variable given the complex topography, ranging from 1,600 mm along parts of the eastern escarpment edge to less than 600 mm in the rain-shadow zone, 20 km west of the escarpment (Sene et al. 1998). Mean annual temperatures along the escarpment summits are 5–6 °C and seasonally range from ~10 °C (NJF) to ~0 °C (JJ) (Grab 2013). Frost days vary from ~30 to 40 days in the foothills to >120 days in high mountain valleys (Grab 2013), thus supporting alpine flora and a marginally periglacial physical setting. Cooler climatic conditions during Quaternary glacial phases, however, were more strongly associated with land surface instability under periglacial and glacionival conditions. This chapter summarizes (1) the main physical properties of the Drakensberg escarpment which have strongly influenced its landscape development, (2) distinctive landforms of the Drakensberg shaped by weathering, periglacial/glacial, mass movement and fluvial processes, in particular during the Quaternary, and (3) some of the challenges to landscape geomorphology, geo- and bioconservation in the light of present-day climate change and human activity.

6.2 Geological Setting and Pre-Quaternary Landscape Evolution

The rock types and thus long-term geological history and patterns of weathering and denudation of the Great Escarpment vary along its length. Sedimentary rocks of the Karoo Supergroup are overlain by the Drakensberg Group volcanic lavas. The lavas produced a ~1,500-m-thick basaltic capping along the Drakensberg escarpment (Duncan and Marsh 2006) and formed almost horizontally layered basalt flows (represented as scarp faces) individually 3–8 m thick (Grab et al. 2005). The lithological contrasts, age and formation mechanisms of adjacent Karoo Supergroup rocks have exerted a strong control on their subsequent weathering rates and processes, which in turn have implications for contemporary landforms and long-term landscape evolution.

Following Karoo Supergroup (and Drakensberg Group) formation, denudation and land surface erosion of the Great Escarpment region took place episodically throughout the Cenozoic. Evidence for the degree to which erosion has affected the land surface comes from exhumed igneous rocks, accordant summits that correspond to preserved palaeo-surfaces and apatite fission track thermochonometric dating that suggests erosion rates of 95 ± 43 m/Myr between 91 and 69 Myr (Brown et al. 2002). Phases of tectonic uplift were followed by land surface denudation that purportedly formed regional-scale planation surfaces, including the African Surface (Late Cretaceous), Post-African I Surface (Early Miocene) and Post-African II Surface (Pliocene) (Partridge and Maud 2000). More specifically, using ^{36}Cl exposure dating, Fleming et al. (1999) calculated summit erosion rates of 6 m/Myr and escarpment retreat rates of 50–95 m/Myr for the Drakensberg sector. However, there also would have been high spatial and temporal variability of denudation rates, for which there is little evidence. As such, more recent research has focused on the Late Quaternary and Holocene in the Drakensberg escarpment region.

6.3 Landscape and Climate Dynamics During the Quaternary and Holocene

Across Africa as a whole, coeval shifts in temperature and precipitation conditions caused by changes in atmospheric circulation took place during the Quaternary (Tyson 1999) and Holocene (Roberts et al. 2013). There is uncertainty in the magnitude of temperature and precipitation changes at this time, largely because these values were likely affected by distance from moisture source and orographic effects, as they are today. Pollen and speleothem records suggest a temperature decrease during the Last Glacial Maximum (LGM; around 20,000 years ago) of 5–7 °C relative to present. The high relief in the Drakensberg region likely means that there was significant climate variability between adjacent summit and valley sites over very short spatial scales. This means that regional climate patterns may not always describe microscale climates.

At a landscape scale (<100 km^2) along the Drakensberg escarpment, spatial and temporal patterns of Quaternary and Holocene changes are as yet not fully understood due to a lack of high-resolution geo- and biological proxy data. However, it is likely that, as can be seen today, the Drakensberg experienced significant climatic differences between lowland and upland sites, and on slopes of different aspect. This makes it difficult to generalize the climate records from any one mountain site to adjacent areas. The topographic setting to the west of the steep Great Escarpment is one of undulating but generally altitudinally declining mountain slopes. Here, the preservation of deep sedimentary sequences in valley heads (hollows >5 m deep) and as colluvial mantles flanking high alpine slopes (up to 13 m deep) has yielded the oldest ^{14}C ages of 13,500 and 43,000 years ago, respectively (Marker 1995; Grab and Mills 2011). The contemporary semi-arid climate west of the Great Escarpment, together with overgrazing, has consequently promoted gully erosion into these sediments, exposing the sedimentary sequences below (Fig. 6.3).

Fig. 6.3 Gully (donga) erosion adjacent to the Great Escarpment, revealing a sediment succession that records Late Pleistocene to Holocene climatic events (*Photograph* S. Grab)

Despite the excellent potential these sediments offer to understanding the alpine Holocene environmental history (Roberts et al. 2013); knowledge is still limited to broadly warmer/wetter conditions between 15,500 and 9,000 years ago, a drier/colder period between 9,000 and 5,000 years ago, warmer/wetter times between 5,000 and 1,000 years ago and relatively dry conditions during the past 1,000 years (Marker 1995).

It is likely that temperature departures were enhanced along the Great Escarpment, relative to surrounding lower regions, during cold periods due to the cooling effects of greater snow cover and thus higher albedo. Apart from likely sporadic permafrost on the highest summits during the LGM (based on the presence of large sorted circles) (Grab 2000), permanent snow/ice or permafrost was most unlikely at any stage during the Holocene. However, solifluction and the formation of stone/turf-banked lobes associated with seasonally frozen ground was likely prominent during colder/drier periods within the Holocene, such as the Little Ice Age, which was regionally consistent in its climatic reach across southern Africa, with temperatures 2–3 °C lower than present (Holmgren et al. 1999).

6.4 Description of Drakensberg Geomorphology

In order to illustrate different types of geomorphic features formed during the Quaternary and Holocene in the Drakensberg escarpment region, we use the Sani Pass upland area of easternmost Lesotho as an example, an area known to have been affected by significant climate changes during the Quaternary and Holocene (Marker 1994).

6.4.1 Rock Weathering Phenomena

Rock weathering throughout the Cenozoic has resulted in a range of small landforms that reflect the interplay between physical (mechanical), chemical and biological weathering and erosion processes, which are largely driven by climate. Evidence for weathering over long time periods comes from the presence of (1) weathered igneous rocks to 25 m depth below the surface (Bell and Haskins 1997), leaving corestones with prominent attributes of exfoliation (Fig. 6.4a); (2) the formation of pseudokarst weathering pits on exposed basalt and sandstone rock surfaces, comprising circular hollows on rock surfaces that are sometimes filled with water (Grab et al. 2011) (Fig. 6.4b); (3) fractured bedrock scarp edges in which fractures extend several metres in depth and which are likely products of mechanical (freeze-thaw; thermoclastis, wetting and drying cycles) and chemical weathering (Fig. 6.4c); and (4) freshly exposed and angular,

Fig. 6.4 Weathering features in the high Drakensberg: **a** Corestones produced by subsurface chemical weathering; **b** pseudokarst surface weathering pits found on basalt outcrops (tape measure is extended to 50 cm); **c** fractures developed around basalt scarp edges; **d** impacts on basalt bedrock produced through lightning strikes. Handheld GPS for scale is sitting on boulder surface on middle right (*Photographs* J. Knight)

fractured bedrock material that has been removed from the blasting effect of lightning strikes (Fig. 6.4d) (Knight and Grab 2014). The consequence of such land surface weathering and downslope sediment transport is the formation of relatively bare upland surfaces with scattered remanié boulders. Cosmogenic dating of land surfaces along the Great Escarpment can be used to calculate rates of denudation over the Late Quaternary and Holocene and shows strong geologic control on denudation patterns and at rates of 1–3 m/Ma (Kounov et al. 2007).

6.4.2 Periglacial and Glacial Features

Higher altitudes (>3,200 m a.s.l) of the Drakensberg escarpment are today only marginally periglacial in character. In these areas, earth hummocks (thúfur) and microscale sorted stone circles have been observed (Fig. 6.5a), suggesting that the contemporary alpine climate can sustain these relatively small-scale phenomena. However, it is likely that glacionival conditions (permanent snowbanks and niche/cirque glaciers) existed at a few topographically favourable localities during the Late Quaternary. Evidence for this comes from the presence of ridges and banks of sediment found on high-altitude, south-facing slopes of the Leqooa Valley in particular, interpreted as ice-marginal moraines (Mills et al. 2009, 2012) (Fig. 6.5b). More commonly, evidence for past periglacial activity has been observed in the Drakensberg, including large (>1 m diameter) stone circles, stone-banked lobes, solifluction lobes and block streams. These larger-scale landforms which are not presently active suggest development during cooler and probably more humid periods, such as the last glaciation and possibly during the Little Ice Age; yet evidence as to their exact age remains lacking.

6.4.3 Mass Movements

Associated with Quaternary and Holocene climate changes, and triggered in part by the near-vertical cliffs along the escarpment edge, are a range of mass movement features to

Fig. 6.5 Periglacial and glacial landforms in the high Drakensberg: **a** Earth hummocks (thúfur); **b** lateral glacial moraine ridge (*Photographs* S. Grab)

the east of the Great Escarpment. The high escarpment edge is exceptionally exposed to high lightning frequency where rock-exposed surfaces are regularly blasted and weakened through enhanced fracturing (Knight and Grab 2014). The majority of rockfalls along the Great Escarpment as a whole originate near the summit edge where mechanical fracturing is most enhanced, and these subsequently progress down the escarpment wall, sometimes developing into rock slides or triggering dry debris flows on the lower flanks (see Fig. 6.6a, b for active rockfall events). Below the escarpment wall, where gradients are less steep and regolith cover is generally >0.5 m, mass movements are dominated by surficial

Fig. 6.6 Mass movement landforms in the high Drakensberg escarpment region: **a** evidence for a rockfall caused by toppling. Person for scale to the left of the rockfall; **b** unconsolidated rockfall in action down the Drakensberg escarpment edge (Photograph from http://www.vertical-endeavour.com); **c** rotational landslide scar (12 m wide) developed in valley-fill sediments within a Drakensberg valley (*Photographs* (**a**) and (**c**) by S. Grab)

Fig. 6.7 Fluvial landforms in the high Drakensberg escarpment region: **a** waterfall into an incised bedrock canyon (*Photograph* S. Grab); **b** low-relief river channel with a mixed bedrock and gravel substrate (*Photograph* J. Knight)

slides and flows, but less frequent translational and rotational landslides and mudflows also occur (Fig. 6.6c). In some instances, slides with a prominent shear zone develop into flows as they incorporate sediment and water within the transportation zone. Rotational slides are often developed through the basalt into the underlying Clarens Formation and tend to fail through a combination of infiltrating rainwater and brittle rock fracture. Large landslide events around the Drakensberg margins, dating most likely to the Late Quaternary/Early Holocene, have been instrumental in blocking some river valleys and diverting drainage. Over millennial time scales, soil erosion and slope sediment supply into mountain valleys are mainly controlled by climate, and dated periods of highest sediment supply are associated with periods of increased humidity and thus subsurface moisture availability (Grab and Mills 2011).

6.4.4 Fluvial Landforms

Over long time scales, tectonic uplift of the Great Escarpment region has been associated with changing patterns of river headcutting and drainage diversion (Roberts and White 2010) (Fig. 6.7a). Rivers in mountain valleys regularly experience flash floods during the wet season and have highly episodic patterns of bedload transport and are consequently characterized by boulder- and gravel-dominated beds at the base of the Great Escarpment and bedrock-dominated channels to the west of the escarpment (Fig. 6.7b). Terraces and abandoned (relict) bars and small floodplains are commonly observed within valleys, but these largely result from fluvial reworking of slope-derived materials, and the spatial pattern of slope sediment supply is highly variable with some valley sides and floors very extensively covered with deep sediment, whereas adjacent valleys may be very sediment poor. Within and adjacent to river channels, elongated gravel bars are formed, which shows the importance of episodic high-energy flood events as a control on patterns of sediment erosion and aggradation within mountain catchments (Grenfell et al. 2008). Furthermore, these events impact directly on patterns of geohazards downstream.

6.5 Present Management Issues in the Drakensberg Escarpment Region

The Drakensberg escarpment region forms part of the uKhahlamba/Drakensberg Park World Heritage Site (inscribed 2000), identified on the basis of its interconnections between the physical (outstanding topography, high biodiversity and endemic species) and human landscapes (San rock art) (see also Chaps. 2 and 19). This area is also protected as part of the Maluti-Ukhahlamba/Drakensberg Transfrontier Conservation Area (established 2001). However, overgrazing and related soil erosion is the most pressing issue in the sustainability of mountain landscapes along the Great Escarpment generally, and the high Drakensberg sector in particular. Furthermore, the construction of new roads and dams with associated quarrying, leaving many wetlands and mountain slopes permanently scarred, is an ever-growing concern (Fig. 6.8). Additional planned developments include the building of new tourist lodges, construction of wind turbines, a cableway from the foothills to the escarpment summit and additional diamond mining excavations—all of which are likely to negatively impact on the aesthetic landscape appeal of the Great Escarpment. Finally, ongoing climate change and associated human–landscape adaptation measures (e.g. grazing wetlands and steeper slopes, more regular grass burning) adversely affect

Fig. 6.8 The role of human activity inducing geomorphic change, Drakensberg escarpment region: a new roadway being opened up by the use of dynamite and leading to increased slope instability and associated mountain geohazards (*Photograph* S. Grab)

land surface stability and geohazards. These activities pose issues for maintaining biodiversity and cultural heritage (e.g. Clark et al. 2011; Duval and Smith 2013).

6.6 Summary

The geomorphological evolution of the Drakensberg escarpment is closely related to initial volcanic and plate tectonic events and subsequent patterns of climate and environmental changes throughout the Cenozoic. The relatively cold and at times wet climate during the Late Quaternary has left a host of prominent palaeo-landforms which have assisted the construction of a crude palaeo-climate history for the high Drakensberg—a limiting factor being the lack of an accurate and high-resolution dating chronology. Changes in macroscale denudation rates has shaped the nature of the land surface and provided the source materials for subsequent ongoing microscale weathering, and slope sediment reworking and transport within slope, mass movement, periglacial and fluvial systems. More recently, human-induced activity has left a prominent and irreversible imprint on this landscape through overgrazing and grassland burning, causing widespread gully erosion and exposing the surface to increased contemporary aeolian processes and associated landform development (turf exfoliation, pan formation). The Drakensberg escarpment thus provides a good example of the strong interrelationship between geology, climate and landscape evolution.

Acknowledgments We thank Stephanie Mills and Wishart Mitchell for comments on a previous version of this chapter.

References

Bell FG, Haskins DR (1997) A geotechnical overview of Katse dam and transfer tunnel, Lesotho, with a note on basalt durability. Eng Geol 46:175–198

Brown RW, Summerfield MA, Gleadow AJW (2002) Denudational history along a transect across the Drakensberg Escarpment of southern Africa derived from apatite fission track thermochronology. J Geophys Res 107(B12). doi:10.1029/2001JB000745

Clark VR, Barker N, Mucina L (2011) Taking the scenic route—the southern Great Escarpment (South Africa) as part of the Cape to Cairo floristic highway. Plant Ecol Divers 4:313–328

Duncan AR, Marsh JS (2006) The Karoo igneous province. In: Johnson MR, Anhaeusser CR, Thomas RJ (eds) The geology of South Africa. Geological Society of South Africa, Johannesburg/Council for Geoscience, Pretoria, pp 501–520

Duval M, Smith B (2013) Rock art tourism in the uKhahlamba/ Drakensberg world heritage site: obstacles to the development of sustainable tourism. J Sust Tour 21:134–153

Fleming A, Summerfield MA, Stone JO, Fifield LK, Cresswell RG (1999) Denudation rates for the southern Drakensberg escarpment, SE Africa, derived from in-situ-produced cosmogenic ^{36}Cl: initial results. J Geol Soc Lond 156:209–212

Grab S (2000) Stone-banked lobes and environmental implications, high Drakensberg, southern Africa. Permafrost Periglac Process 11:177–187

Grab SW (2013) Fine-scale variations of near-surface-temperature lapse rates in the high Drakensberg Escarpment, South Africa: environmental implications. Arct Antarct Alp Res 45:500–514

Grab SW, Mills SC (2011) Late quaternary slope environments in the upper Sehonghong Valley, eastern Lesotho. Proc Geol Assoc 122:179–185

Grab S, van Zyl C, Mulder N (2005) Controls on basalt terrace formation in the eastern Lesotho highlands. Geomorphology 67:473–485

Grab SW, Goudie AS, Viles HA, Webb N (2011) Sandstone geomorphology of the Golden Gate Highlands National Park, South Africa, in a global context. Koedoe 53(1) Art. #985, 14 pp. doi:10.4102/koedoe.v53i1.985

Grenfell MC, Ellery W, Grenfell SE (2008) Tributary valley impoundment by trunk river floodplain development: a case study from the KwaZulu-Natal Drakensberg foothills, eastern South Africa. Earth Surf Proc Land 33:2029–2044

Holmgren K, Karlén W, Lauritzen SE, Lee-Thorp JA, Partridge TC, Piketh S, Repinski P, Stevenson C, Svanered O, Tyson PD (1999) A 3000-year high-resolution stalagmite-based record of palaeoclimate for northeastern South Africa. Holocene 9:295–309

King LC (1953) Canons of landscape evolution. Bull Geol Soc Am 64:721–762

Knight J, Grab SW (2014) Lightning as a geomorphic agent on mountain summits: evidence from southern Africa. Geomorphology 204:61–70

Kounov A, Niedermann S, de Wit MJ, Viola G, Andreoli M, Erzinger J (2007) Present denudation rates at selected sections of the South African escarpment and the elevated continental interior based on cosmogenic ^{3}He and ^{21}Ne. S Afr J Geol 110:235–248

Marker ME (1994) Sedimentary sequences at Sani Top, Lesotho highlands, southern Africa. Holocene 4:406–412

Marker ME (1995) Late quaternary environmental implications from sedimentary sequences at two high altitude Lesotho sites. S Afr J Sci 91:294–298

Mills SC, Grab SW, Carr SJ (2009) Recognition and palaeoclimatic implications of late quaternary niche glaciations in eastern Lesotho. J Quat Sci 24:647–663

Mills SC, Grab SW, Rea BR, Carr SJ, Farrow A (2012) Shifting westerlies and precipitation patterns during the Late Pleistocene in southern Africa determined using glacier reconstruction and mass balance modelling. Quat Sci Rev 55:145–159

Moore A, Blenkinsop T, Cotterill F (2009) Southern African topography and erosion history: plumes or plate tectonics? Terra Nova 21:310–315

Partridge TC, Maud RR (2000) Macro-scale geomorphic evolution of southern Africa. In: Partridge TC, Maud RR (eds) The Cenozoic of Southern Africa. Oxford University Press, Oxford, pp 3–18

Partridge TC, Dollar ESJ, Moolman J, Dollar LH (2010) The geomorphic provinces of South Africa, Lesotho and Swaziland: a physiographic subdivision for earth and environmental scientists. Trans R Soc S Afr 65:1–47

Roberts GG, White N (2010) Estimating uplift rate histories from river profiles using African examples. J Geophys Res 115:B02406. doi:10.1029/2009JB006692

Roberts P, Lee-Thorp JA, Mitchell PJ, Arthur C (2013) Stable carbon isotopic evidence for climate change across the late Pleistocene to early Holocene from Lesotho, southern Africa. J Quat Sci 28: 360–369

Sene KJ, Jones DA, Meigh JR, Farquharson FAK (1998) Rainfall and flow variations in the Lesotho Highlands. Int J Climatol 18: 329–345

Tyson PD (1999) Late-quaternary and Holocene palaeoclimates of southern Africa: a synthesis. S Afr J Geol 102:335–349

Rocky Shorelines of the Wild Coast

7

Jasper Knight and Stefan Grab

Abstract

The bedrock-dominated coastline along the northern part of Eastern Cape, commonly known as the 'Wild Coast', contains a range of coastal erosional and depositional landforms developed in bedrock and unconsolidated sediments. Typical erosional landforms found along this coast include cliffs, shore platforms and relict (fossil) clifflines, sea arches and stacks, and tafoni weathering forms. Depositional features include beachrock, aeolianite, coastal sand dunes, sandy beaches and barriers across tidal inlets. These erosional and depositional landforms are largely a product of climate and sea-level changes during the Quaternary and Holocene. The aesthetically pleasing landscapes of the Wild Coast are a significant tourist attraction.

Keywords

Quaternary • Holocene • Tafoni • Bedrock platform • Aeolianite • Sea-level change

7.1 Introduction

The Wild Coast refers informally to the part of the Eastern Cape Province coastline that extends approximately between the Great Kei River to the south and the Umtamvuna River to the north ($\sim$320 km distance) (Fig. 7.1). The area, also known as the Pondoland coast, has a steep adjacent continental shelf and is characterized by a predominance of near-vertical coastal cliffs and steep-sided rocky inlets terminating at the shoreline. The hinterland receives relatively high annual rainfall and resulting rivers exit into the Indian Ocean through impressive fluvially incised gorges, waterfalls and lagoons. The Wild Coast can be broadly distinguished from adjacent areas because of its relatively linear, southeast-facing coastline, and there is a strong geologic control on the range and geomorphic properties of both erosional and depositional landforms that are found along it. Many tourist 'hot spots' along the Wild Coast coincide with coastal landforms such as 'the Gates' (sandstone cliffs either side of the Mzimvubu River mouth), 'Hole in the Wall' (a rock arch), 'Waterfall Bluff' and the 'Morgan Bay cliffs' (Fig. 7.1). Apart from its scenery, the coastline also contains endemic flora and fauna and has been identified as a transition zone between the subtropical biogeographic province to the north and the warm-temperate province to the south (Turpie et al. 2000).

7.2 Geologic Setting

The northern portion of the Wild Coast between Port St Johns and Port Edward is underlain by the Palaeozoic Msikaba Formation (Natal Group) which comprises a sequence dominated by sandstones and associated mudstones. The central and southern part of this coastline is underlain by sedimentary rocks of the Karoo Supergroup, mainly sandstones and mudstones (Ecca Group, Adelaide Subgroup), although glacigenic diamictites of the Dwyka Group also outcrop along the coast in some places (Johnson et al. 2006) (Fig. 7.1). Unlike other areas of the east coast of South Africa, faulting through the Wild Coast region has

J. Knight (✉) · S. Grab
University of the Witwatersrand, Johannesburg, South Africa
e-mail: jasper.knight@wits.ac.za

S. Grab and J. Knight (eds.), *Landscapes and Landforms of South Africa*,
World Geomorphological Landscapes, DOI 10.1007/978-3-319-03560-4_7,

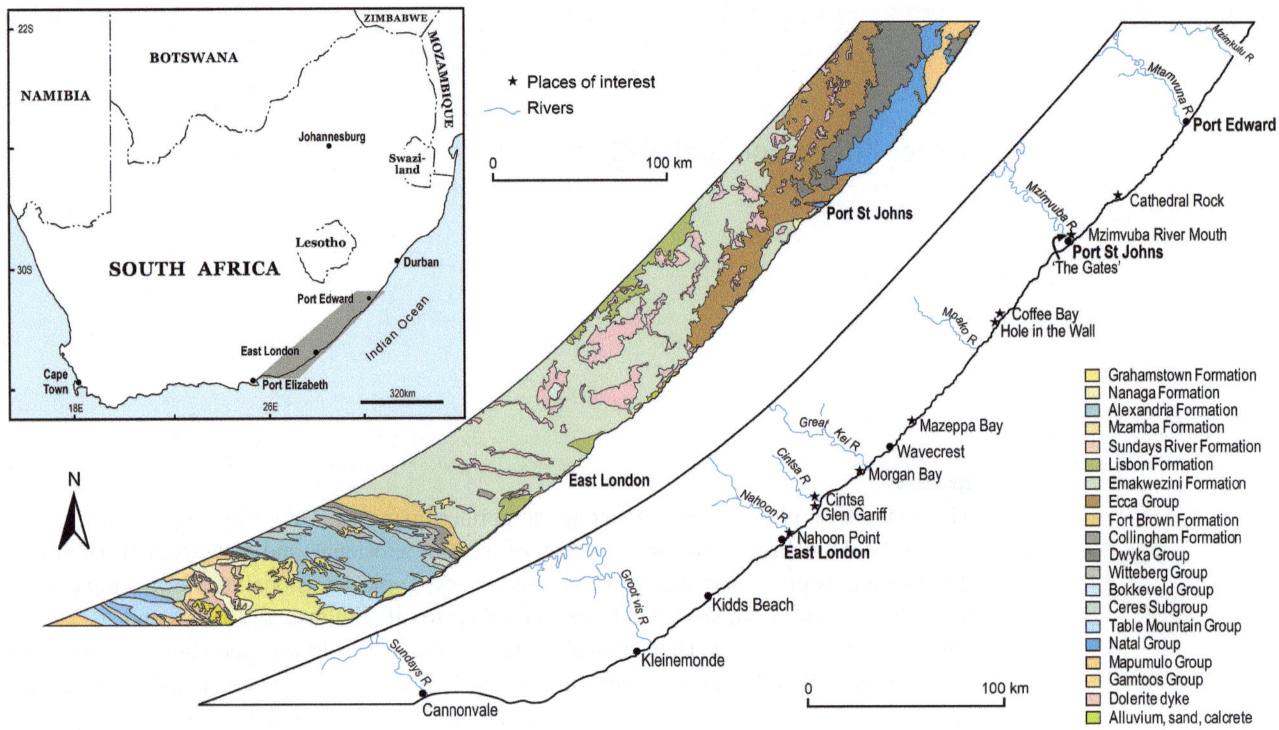

Fig. 7.1 Location and geology of the Wild Coast and sites named in the text

brought the Beaufort Group sediments, in particular the Molteno Formation, to sea level. Thus, several headlands are mainly of Molteno sandstone, capped by Lower Cretaceous sedimentary rocks. Some of the stacks and arches along the coast appear to be a product of Karoo dolerite intrusions.

The varied lithologies that meet the coast exhibit high densities of joints and bedding planes that can be exploited by coastal erosion; thus, erosional landforms predominate. As a consequence of both the shearing of the Falklands microplate during the break-up of Gondwana (at 135 Ma) (Watkeys 2006), and the presence of hard rocks, a steep nearshore zone exists with an extremely narrow continental shelf. Deep water is therefore found close to shore and the steep nearshore gradient means that large waves can break near to the shoreline (e.g. Dingle and Scrutton 1974). An exception to this is within isolated, shallower sandy embayments which are often backed by unconsolidated sand dunes. Because of these physical properties of the coast and its exposure to the Southern Ocean, the wave climate of the east coast of South Africa is relatively vigorous, with a mean significant wave height of 1.65 m (Corbella and Stretch 2012). In some areas, the presence of large shoreline boulders that have moved over historic time periods attests to much higher (>11 m) wave activity under storm conditions (Salzmann and Green 2012). This coastline also has a microtidal regime, which is important because it has implications for the vertical range over which tidal energy, and thus, erosional potential is expended.

The geology, climate and wave regime of the region also influence coastal geomorphology in more subtle ways. For example, bedrock control on the direction and sinuosity of rivers in the adjacent hinterland has resulted in very narrow and bedrock-constrained tidal inlets along this coast (Cooper 1993, 2001). The relatively steep shoreface has meant that extensive intertidal platforms are not found continuously along the Wild Coast, but those that lie adjacent to inlet mouths may have constrained river-mouth positions during periods of lower than present sea level, such as during the last glacial period ~29,000–14,000 year BP) and into the Early Holocene, when river mouths extended farther seaward (Green and Garlick 2011). It is likely that rivers during this glacioeustatic lowstand provided the shelf sediment that was subsequently reworked into sandy embayments and large sand dunes during sea-level rise. This illustrates the interplay between different controls on the formation and dynamics of landforms along the Wild Coast during the Quaternary and Holocene.

In detail, this chapter describes (1) the erosional and (2) depositional landforms of the Wild Coast; (3) their relationship to sea-level change during the Quaternary and Holocene; and (4) present environmental issues affecting this coast. The interrelationship between the development of coastal landforms and climatic driving factors, including sea-level change and wave processes, illustrates the dynamic nature of rocky coasts and their evolution over different spatial and temporal scales. Some of these landforms and sites are discussed in more detail by Illgner (2008).

7.3 Erosional Features

7.3.1 Cliffs and Caves

Cliffs are present on most headlands along this coast, such as 'the Gates' at Port St Johns, and are up to 30 m high. Usually these cliffs are steep-faced (Fig. 7.2a) but fossil clifflines located at different levels above present sea level are often found at the back of embayments or around headland margins where they correspond to previous eustatic sea-level positions (Marker 1987). These fossil clifflines and associated beaches or platforms are sometimes associated with fossil caves, such as at Coffee Bay where a cave at +4.5 m was incised by high sea levels and contains marine shells dated to the last interglacial period (Ramsay and Cooper 2002).

7.3.2 Bedrock Shore Platforms

Bedrock shore platforms within the intertidal zone are well exposed in the absence of surficial sand cover (Fig. 7.2b). These platforms are usually planar, thus corresponding to the zone of greatest tide-driven abrasion, are laterally extensive and have a micromorphology that is strongly affected by geologic structure including the disposition of joints and bedding planes (Truswell 1972). Sedimentary rocks of the Adelaide Subgroup show a shallow northerly dip which gives rise to an alongshore corrugated pattern, in which detached boulders are captured within depressions (Fig. 7.2c).

7.3.3 Arches and Stacks

Erosion of rocky coasts by wave undercutting can result in the formation of arches and stacks. Near Coffee Bay, the Hole in the Wall is an arch developed within a small offshore island, larger than a single rock stack (Fig. 7.2d). This landform reflects progressive focusing of incoming wave energy by refraction, leading to enhanced undercutting around mean tide level and subsequent cliff collapse. The adjacent Mpako River may also have helped widen the arch through tidal pumping (Fig. 7.2e). Arch collapse through continued coastal erosion, especially where bedrock is highly jointed, is clearly shown in Cathedral Rock, an offshore rock stack which contains several basal arches that have exploited bedrock jointing patterns (Fig. 7.2f). Cathedral Rock is highly susceptible to future erosional undercutting and cliff collapse.

7.3.4 Small-Scale Erosion Features

Wave processes within the intertidal zone, associated with turbulent transport of sandy sediment by uprushing waves, can cause abrasion and undercutting of bedrock outcrops (e.g. Fig. 7.2g). Consequently, the downwasting of rock surfaces which, over time, generates loose sediment and forms rock outcrops with intricate micro-scale erosional features. This is particularly prominent along the Wild Coast given the high joint densities of many rock outcrops. Tafoni (honeycomb weathering) is well represented at many locations (e.g. Fig. 7.2h). It is formed by a range of physical and biological weathering processes and is particularly associated with coastal settings where salt spray and bacteria/algae growth are common (McBride and Picard 2004). As such, tafoni develops above high tide positions on platforms or in surrounding cliffs.

7.4 Depositional Features

7.4.1 Tidal Inlets

Tidal inlets and microtidal estuaries are common along the Eastern Cape and KwaZulu-Natal coast, such as the mouths of the Sundays, Kei, Mzimvubu and Mzimkulu rivers (Fig. 7.3). These inlets are all narrow, strongly geologically controlled by rock-cut river channels and constraining rock headlands at river mouths, have a small tidal prism and are most commonly wholly or partially barred by sandy barriers. The properties and morphodynamics of these inlet mouths is strongly influenced by episodic river floods, which open inlet mouths, and ocean storms which cause overwash events, barrier rollover and inlet closure (Cooper 1993, 2001). Most tidal inlets have progressively infilled over the Holocene, but the lagoonal back-barriers formed by this process are often associated with mangroves and other important ecosystems.

7.4.2 Sandy Beaches and Sand Dunes

Sandy and to a lesser extent boulder beaches are found within bedrock embayments along the Wild Coast and are usually associated with coastal sand dunes (Fig. 7.4a, b). Compared to other coastal stretches around South Africa, these beaches are poorly known both in terms of their properties and dynamic behaviour. However, seasonal variations in wave climate are likely to cause cyclic patterns of steepening and flattening of the beach profile and the formation of ridges or berms during winter storm events. Coastal sand dunes are present extensively along this coastline and can take several geomorphic forms that represent the interplay between sediment availability, coastline geometry and direction of prevailing winds (Illenberger and Burkinshaw 2008). Along the Wild Coast, dunes can form a thin coastal fringe, such as near Mazeppa Bay; or as

Fig. 7.2 Photographs of typical erosional landforms of the Wild Coast. **a** Cliffed headland at Morgan Bay with adjacent rock platform, **b** intertidal platform at Great Kei River mouth, **c** geologically controlled morphology of a rock platform at Morgan Bay, showing boulders captured within topographic lows, **d** coastal arch at the Hole in the Wall, **e** oblique view of Hole in the Wall showing its position relative to the mouth of the Mpako River (*left*), **f** view of Cathedral Rock looking onshore, **g** wave undercutting of sandstone outcrop at Kidds Beach, **h** tafoni weathering in sandstone at Morgan Bay (*Photograph* **d** courtesy of Brian Mallinson, *Photograph* **e** courtesy of www.wilderness.co.za, *Photograph* **f** courtesy of Debbie Smith, Offshore Africa Port St John's and all other *Photographs* by J. Knight)

Fig. 7.3 Oblique air photograph of the river mouth and estuary area of Sundays River showing a prominent flood delta (*Photograph* courtesy of Werner Illenberger)

Fig. 7.4 Photographs of typical depositional landforms of the Wild Coast, **a** sandy embayment beach and backing sand dunes which are partially vegetated, Kidds Beach, **b** mixed sand and boulder beach, Great Kei River mouth, **c** cross-bedded sandstones within aeolianite located around 4 m above sea level at Nahoon Point, **d** eroded beachrock platform at St Lucia Bay, outside of the Wild Coast region, but illustrating the main properties of beachrock (*Photographs* J. Knight)

transgressive or landward-migrating dunes such as at Wavecrest; or adjacent to the mouths of coastal inlets where they formed nested sets on either side of the mouth, such as adjacent to the Cintsa River. Dunes can form cuspate forelands such as between Morgan Bay and the Great Kei River mouth where the triangular-shaped foreland is over 3 km wide and 1.4 km in accretionary extent. In other places, coastal dunes are located on top of bedrock platforms, such as at Glen Gariff.

7.4.3 Aeolianite and Beachrock

Aeolianite and beachrock are formed when unconsolidated dune and beach sand grains, respectively, become cemented together, usually by carbonates, as a result of the supersaturation and then degassing of CO_2 from warm (>65 °C) circulating groundwater. The formation of such cemented rocks requires vigorous subsurface water circulation, which can take place in particular within coarse sands in the upper intertidal zone. Here, beachrock can form where fresh groundwater and marine water are moved by wave pumping, and aeolianite can form as a result of groundwater seepage and/or high evaporation rates from sand dunes. Both aeolianite and beachrock are relatively common along the Wild Coast and in adjacent areas because of the relatively warm conditions, both at the present time and during past interglacials (Cooper 1991; Cawthra and Uken 2012); and the presence of underlying impermeable bedrock surfaces may have favoured enhanced groundwater flow and CO_2 degassing. Aeolianite, forming 'fossil dunes', can reach some tens of metres in thickness and preserve cross-bedding indicative of past wind direction (Fig. 7.4c). Aeolianite at Nahoon Point near East London is dated to ~124,000 years BP and is particularly important because it contains contemporary human footprints (Jacobs and Roberts 2009). Beachrock (Fig. 7.4d) generally reflects approximate sea-level position and, because of its cemented nature, can affect subsequent coastal evolution (Cooper 1991). Small-scale erosional features including tafoni are commonly developed in both aeolianite and beachrock by chemical and physical weathering and faunal borers (Miller and Mason 1994).

Similar coastal chemical processes, but which are mediated by biological activity, also form small-scale depositional structures that are found along the Wild Coast. For example, tufa stromatolites are presently forming in supratidal rock pools at Cape Morgan, near to the Kei River mouth (Smith et al. 2011). These unusual structures result from degassing of carbonate-rich groundwater and represent the unique interplay between geology, water circulation and microbial activity in a coastal setting.

7.5 Relationship of Rocky Coast Landforms to Quaternary and Holocene Sea-Level Changes

Evidence for sea-level changes along rocky coasts during the Quaternary and Holocene comes from fossil erosional platforms, cliffs and beaches, wave-cut notches at the cliff base and caves. These landforms are good palaeo-sea-level indicators because they are formed at specific positions within the tidal frame (Trenhaile 2010) and have been recognized along the Eastern Cape (Marker 1987) and KwaZulu-Natal coasts (McCarthy 1967). Depositional coastal elements such as aeolianites and beachrock have also been used as sea-level indicators along the Eastern Cape and KwaZulu-Natal coasts (Cooper and Flores 1991; Watkeys et al. 1993; Cawthra and Uken 2012). Raised coastal elements along the Wild Coast and in adjacent areas provide evidence for higher than present sea levels during the last interglacial period (~105,000–130,000 years BP) (Cooper and Flores 1991; Ramsay and Cooper 2002). Unlike dominantly depositional coasts where sediments can be easily eroded, rocky coasts can record sea-level changes over longer time periods where sea-level indicators can be preserved in such bedrock hollows, at the back of embayments, or preserved by chemical lithification in the formation of aeolianite and beachrock. As this is the case with the Wild Coast, there is thus great future potential for studies along this coast to inform on processes and patterns of Quaternary and Holocene sea-level change and their sedimentary and biological imprints.

7.6 Environmental Issues Affecting the Wild Coast

Although rocky coasts are often seen as relatively insensitive to ongoing environmental change, the Wild Coast is affected by several environmental issues. Development of tourism and infrastructure has potential to affect coastal morphodynamics such as longshore sediment transport, tidal inlet functioning, sand dune stability and ecosystems, and pollution and waste water dispersal (Watkeys et al. 1993; Palmer et al. 2010). Ongoing climate change associated with global warming is also affecting all aspects of rocky coast environments. For example, over recent decades, an increase in sea level of ~1.5 mm year^{-1} is observed along the eastern South African coast (Jury 2013). Although consistent with global-scale patterns, this rate of sea-level rise is higher than that on the west coast of South Africa because of the presence of the warm Agulhas Current and due to strong onshore winds and more erodible bedrock substrates (Mather et al. 2009). Furthermore, sea-level rise on rocky coasts is generally associated with larger waves being able to penetrate higher up the tidal frame and is therefore likely to result in higher rates of coastal erosion and increased frequency of single events such as cliff collapse. Changes in regional temperature and precipitation regimes can also impact on cliff face stability and rock surface and subsurface weathering. The rocky shoreline of the Wild Coast is an area of high biodiversity and supports specific ecosystems such as mangroves and intertidal rock pools that are dependent on wave, tide and rock surface temperature regimes (Bustamante and Branch 1996). Maintaining and enhancing such biodiversity requires monitoring present habitats, as well as rates of coastline retreat and rocky shore downwearing, in order to ensure that such changes are managed effectively.

7.7 Conclusions

The Wild Coast contains a range of erosional landforms that are typically found along rocky coasts such as cliffs, arches, waterfalls and shore platforms, but also shows considerable evidence for depositional landforms that have helped provide a record of sea level and environmental changes during the Quaternary and Holocene, such as sand dunes, aeolianite and beachrock. As such, the Wild Coast is an amalgam of palaeo (relict) and contemporary landforms that demonstrate the interplay between geology, geomorphology and climate over long timescales. This, together with the rich regional bio- and cultural- heritage, has encouraged ecotourism initiatives such as the spatial development initiative (SDI), which aims to boost infrastructure and local economic development within otherwise impoverished regional communities.

Acknowledgments We thank Andrew Green and Hayley Cawthra for their comments on a previous version of this chapter.

References

Bustamante RH, Branch GM (1996) Large scale patterns and trophic structure of southern African rocky shores: the roles of geographic variation and wave exposure. J Biogeogr 23:339–351

Cawthra H, Uken R (2012) Modern beachrock formation in Durban, KwaZulu-Natal. S Afr J Sci 108:Art#935, 5p. http://dx.doi.org/10.4102/sajs.v108i7/8.935

Cooper JAG (1991) Beachrock formation in low latitude: implications for coastal evolutionary models. Mar Geol 98:145–154

Cooper JAG (1993) Sedimentation in the cliff-bound, microtidal Mtamvuna estuary, South-Africa. Mar Geol 112:237–256

Cooper JAG (2001) Geomorphological variability among microtidal estuaries from the wave-dominated South African coast. Geomorphol 40:99–122

Cooper JAG, Flores RM (1991) Shoreline deposits and diagenesis resulting from 2 late Pleistocene highstands near +5 and +6 meters, Durban, South-Africa. Marine Geol 97:325–343

Corbella S, Stretch DD (2012) The wave climate on the KwaZulu-Natal coast of South Africa. J S Afr Inst Civil Eng 54:45–54

Dingle RV, Scrutton RA (1974) Continental breakup and development of post-Paleozoic sedimentary basins around Southern Africa. Geol Soc Am Bull 85:1467–1474

Green A, Garlick GL (2011) A sequence stratigraphic framework for a narrow, current-swept continental shelf: The Durban Bight, central KwaZulu-Natal, South Africa. J Afr Earth Sci 60:303–314

Illenberger WK, Burkinshaw JR (2008) Coastal dunes and dunefields. In: Lewis CA (ed) Geomorphology of the Eastern Cape. NISC, Grahamstown, pp 85–106

Illgner PM (2008) Coastal processes and associated landforms. In: Lewis CA (ed) Geomorphology of the Eastern Cape. NISC, Grahamstown, pp 63–84

Jacobs Z, Roberts DL (2009) Last Interglacial age for aeolian and marine deposits and the Nahoon fossil human footprints, southeast coast of South Africa. Quatern Geochronol 4:160–169

Johnson MR, van Vuuren CJ, Visser JNJ, Cole DI, Wickens HdeV, Christie ADM, Roberts DL, Brandl G (2006) Sedimentary rocks of the Karoo Supergroup. In: Johnson MR, Anhaeusser CR, Thomas RJ (eds) The Geology of South Africa. Geological Society of South Africa, Johannesburg/Council for Geoscience, Pretoria, pp 461–499

Jury MR (2013) Climate trends in southern Africa. S Afr J Sci 109(1/2): Art. #980, 11 pp, http://dx.doi.org/10.1590/sajs.2013/980

Marker ME (1987) A note on marine benches of the southern Cape. S Afr Tydskr Geol 90:120–123

McBride EF, Picard MD (2004) Origin of honeycombs and related weathering forms in Oligocene Macigno Sandstone, Tuscan coast near Livorno, Italy. Earth Surf Proc Landf 29:713–735

McCarthy MJ (1967) Stratigraphical and sedimentological evidence from the Durban region of major sea-level movements since the Late Tertiary. Trans Geol Soc S Afr 70:135–165

Miller WR, Mason TR (1994) Erosional features of coastal beachrock and aeolianite outcrops in Natal and Zululand, South-Africa. J Coastal Res 10:374–394

Mather AA, Garland GG, Stretch DD (2009) Southern African sea levels: corrections, influences and trends. Afr J Mar Sci 31:145–156

Palmer BJ, McGregor GK, Hill TR, Paterson AW (2010) A spatial assessment of coastal development and land use change in the Eastern Cape, South Africa. S Afr Geogr J 92:117–128

Ramsay PJ, Cooper JAG (2002) Late Quaternary sea-level change in South Africa. Quatern Res 57:82–90

Salzmann L, Green A (2012) Boulder emplacement on a tectonically stable, wave-dominated coastline, Mission Rocks, northern KwaZulu-Natal, South Africa. Mar Geol 323-325:95–106

Smith AM, Andrews JE, Uken R, Thackeray Z, Perissinotto R, Leuci R, Marca-Bell A (2011) Rock pool tufa stromatolites on a modern South African wave-cut platform: partial analogues for Archaean stromatolites? Terra Nova 23:375–381

Trenhaile AS (2010) The effect of Holocene changes in relative sea level on the morphology of rocky coasts. Geomorphol 114:30–41

Truswell JF (1972) Sandstone sheets and related intrusions from Coffee Bay, Transkei, South Africa. J Sed Petrol 42:578–583

Turpie JK, Beckley LE, Katua SM (2000) Biogeography and selection of priority areas for conservation of South African coastal fishes. Biol Cons 92:59–72

Watkeys MK (2006) Gondwana break-up: a South African perspective. In: Johnson MR, Anhaeusser CR, Thomas RJ (eds) The geology of South Africa. Geological Society of South Africa, Pretoria, pp 531–539

Watkeys MK, Mason TR, Goodman PS (1993) The rôle of geology in the development of Maputaland, South Africa. J Afr Earth Sci 16:205–221

The Augrabies Falls Region: A Fluvial Landscape Divided in Flow but Magnificent in Spectacle

8

Stephen Tooth

Abstract

The fluvial landscape of the arid Augrabies Falls region is characterised by a complex of channels, waterfalls and gorges eroded in granitoid bedrock. Anabranches of the Orange River divide and rejoin around stable, predominantlybedrock islands, and many terminate at or near the permanently flowing 50–60 m high Main Falls, with the largest summer floods also activating other normally dry waterfalls. Below the Main Falls, a single channel flows within a deep, narrow gorge up to ~18 km long. A developmental model is outlined, which posits that waterfall retreat and concomitant gorge formation in the initially low-relief valley floor has initiated a wave of erosion that represents a renewed phase oflandscape denudation . The faster retreat of the Main Falls is driving changes to upstream flow patterns. Over a long time scale, incision along anabranches that are tributary to the gorge will combine with gorge sidewall retreat, thereby leading to valley floor dissection.

Keywords

Anabranching river • Bedrock • Gorge • Knickpoint • Waterfall

8.1 Introduction

The Augrabies Falls region, Northern Cape Province (Fig. 8.1a), is located in the valley of the Orange River, the largest westward-flowing river in southern Africa. The region is distinguished by waterfalls ('falls') that occur on some of the numerous anabranching channels ('anabranches') that characterise this river reach (Fig. 8.1b–d). The name 'Augrabies' (sometimes spelt 'Aughrabies' in older sources) appears to be derived from a corruption of the Nama word '*Akoerabis*' or '*Aukoerebis*', meaning 'Place of Great Thunder', 'Place of Great Noise' or 'Place of Loud Noise' (e.g. van der Walt 2000). This is most likely a reference to the river flow that cascades noisily over the Main Falls and into a gorge all year round, with other named and many unnamed falls becoming active during large flood events (Figs. 8.1d and 8.2). The Main Falls and gorge are located in the Augrabies Falls National Park (proclaimed 1966), which now covers ~554 km^2 and incorporates one of the finest South African examples of a desert landscape fashioned in crystalline (igneous and metamorphic) rocks. This landscape also encompasses mountain ranges and a variety of weathering landforms, including bornhardts, nubbins and pediments.

These rocky desert landforms are the main visitor attractions in the National Park but little specific information is provided at the Visitors Centre or on websites, in part because geomorphological research in the region has been limited. This chapter outlines the main fluvial features of the region, thereby contributing to enhanced appreciation of this aspect of South Africa's geoheritage.

8.2 Geographical and Environmental Setting

In the Augrabies Falls region, annual rainfall averages ~125 mm while potential annual evapotranspiration is >2,500 mm (Water Research Commission 1999; SANParks

S. Tooth (✉)
Aberystwyth University, Wales, UK
e-mail: set@aber.ac.uk

S. Grab and J. Knight (eds.), *Landscapes and Landforms of South Africa*,
World Geomorphological Landscapes, DOI 10.1007/978-3-319-03560-4_8,

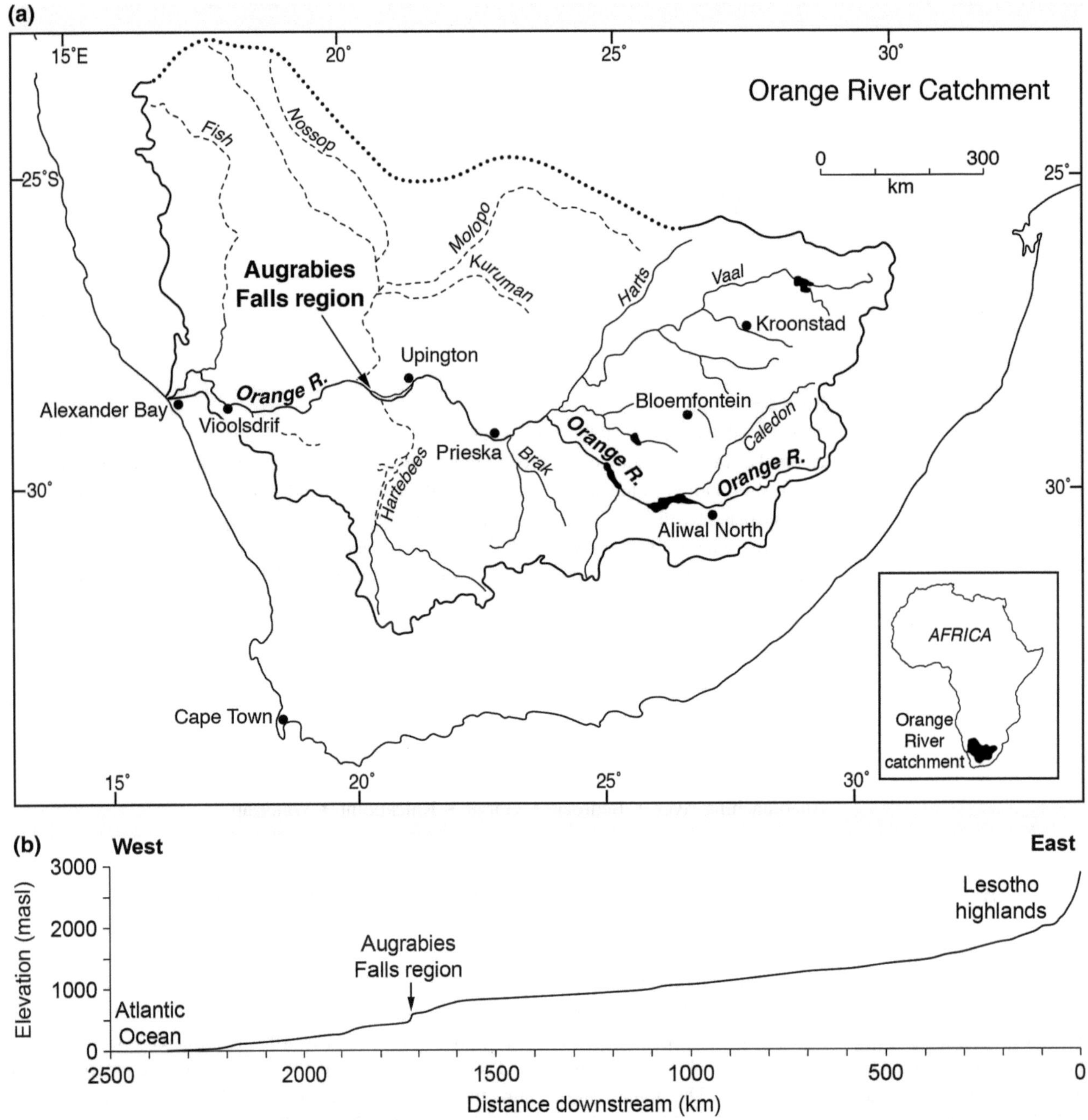

Fig. 8.1 Geographical setting and geomorphological context of the Augrabies Falls region: **a** location within the Orange River catchment; **b** long profile of the Orange River (from 1:250,000 scale topographic maps with 50 m contour intervals), illustrating how the Augrabies Falls region corresponds with a major drop in river bed elevation; **c** aerial photograph (from Google Earth) of the Augrabies Falls region (image centred on 28° 34′ 27″S, 20° 18′ 19″E); **d** map of waterfalls/knickpoints and anabranches covered in c)

2013). Despite this large water deficit, rainfall and snowmelt derived from the upper reaches of the Orange River means that some anabranches maintain low flows all year round (combined winter discharge is typically <100 m^3/s), much of which cascades over the Main Falls. The flow regime is highly variable, however, with larger summer floods commonly in the range 1,000–3,000 m^3/s (e.g. April 2006, February 2010) and occasionally reaching higher values (e.g. nearly 8,000 m^3/s in March 1988—du Plessis et al. 1989; Bremner et al. 1990). These larger floods activate anabranches that supply flow to the other falls and the gorge, and may also inundate other parts of the valley floor. Bedrock geology in the Augrabies Falls region consists mainly of granite gneiss and gneiss (hereafter, 'granitoids') but

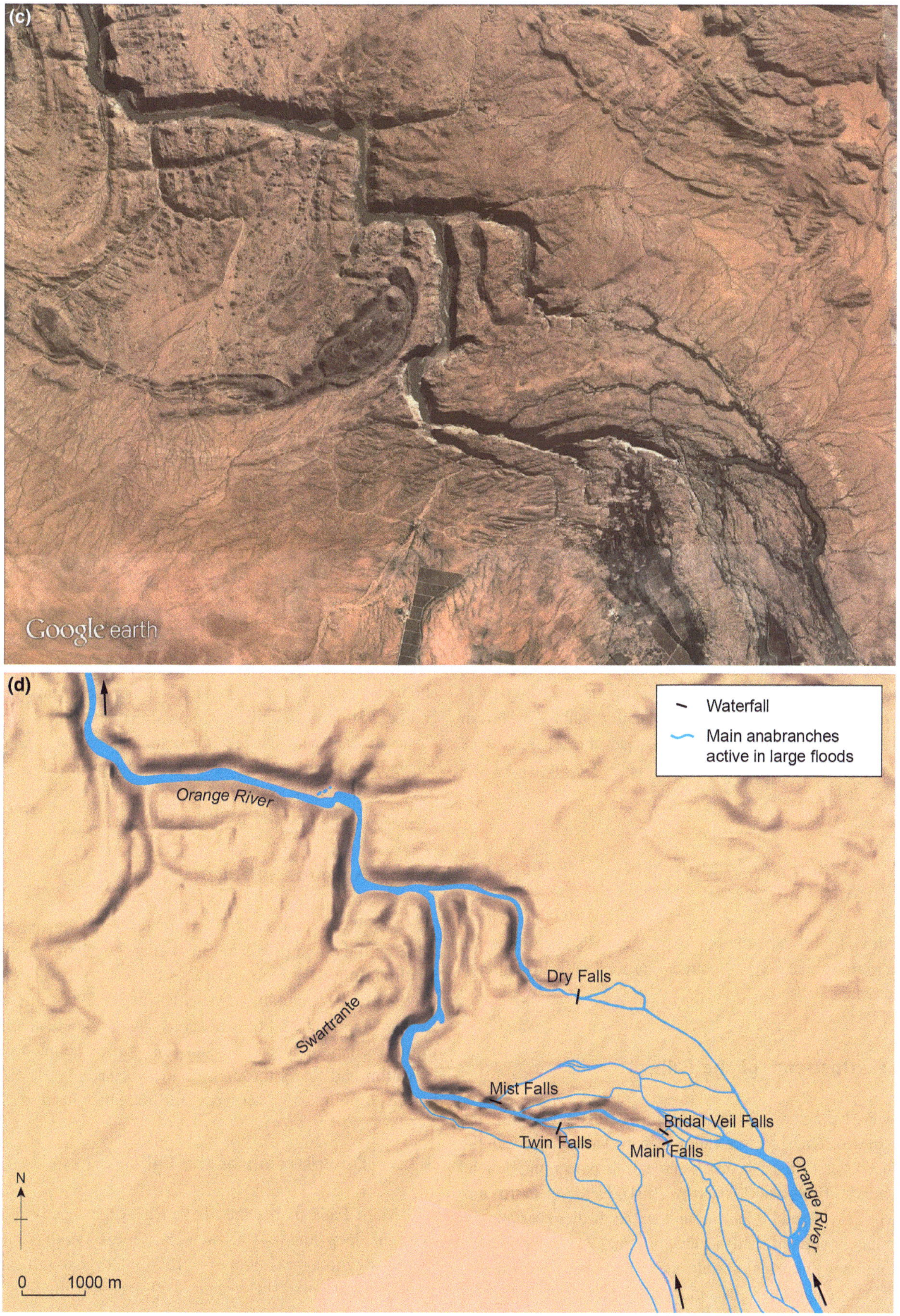

Fig. 8.1 (continued)

Fig. 8.2 Photographs of **a** the Main Falls at Augrabies (view looking downvalley over the lip of the falls); and **b** downstream gorge (view looking upstream, with the Main Falls 4 km distant and out of sight) (*Photographs* S. Tooth)

mafic lithologies crop out locally. Although the region is generally considered to be tectonically stable, a recent (2010–2012) earthquake swarm illustrates the seismic forces still acting on areas of crustal weakness (Council for Geoscience 2013).

8.3 Overview of Landforms and Landscape

The fluvial landscape of the Augrabies Falls region can be divided into three zones: upstream, around and downstream of the Main Falls.

8.3.1 Upstream of the Falls

Between Upington and the Augrabies Falls (150 km), the 2–4 km wide Orange River valley is characterised by one of the world's largest mixed bedrock-alluvial anabranching channel complexes (Tooth and McCarthy 2004). Here, up to 15 anabranches divide and rejoin around stable, bedrock or alluvial (dominantly silt and sand) islands that may be up to ~2 km wide, up to ~15 km long and elevated >10 m above normal river level. In the Augrabies Falls region, alluvium is relatively sparse and most of the anabranches and islands are formed in granitoid bedrock. Structural control by joints, fractures and foliations is clearly evident, with many relatively straight river sections being separated by abrupt bends (Fig. 8.1c, d). These lines of weakness have facilitated the continued division of flow between anabranches, resulting in deep channels and rocky islands.

8.3.2 Around the Falls

The complex of anabranches terminates at the various waterfalls, with the two principal ones being the Main Falls in the south and the Dry Falls located ~2.5 km to the north-west (Fig. 8.1c, d). Anabranch gradients steepen noticeably as the falls are approached. For instance, over a 1.5 km distance, the primary anabranch to the Main Falls drops about 30 m in elevation and flows through a narrow chasm ~10 m deep before plunging ~50–60 m over the lip of the falls (Fig. 8.1b).

8.3.3 Downstream of the Falls

The Main Falls marks the start of a gorge that is initially ~75 m deep and ~100 m wide. Gorge depth increases downvalley to a maximum of ~160 m. Gorge width is more irregular and varies between ~100 and 400 m. Structural control on gorge orientation is clearly evident (Wellington 1955; Slabbert and Malherbe 1983), with straighter sections several kilometres long being separated by abrupt, right-angled bends where the gorge tends to be widest. A number of deeply incised anabranches are tributary to this gorge, including the channel in the ~3.5 km long gorge developed downstream of the Dry Falls (Fig. 8.1d). During large floods, these tributaries transport sand through to boulders. The gorge generally has steep, near-vertical sidewalls, but there is evidence of local mass failure, including toppling of joint-bounded blocks and spalling of massive sheets, which supplies additional coarse debris to the gorge floor (Fig. 8.2b). Some of this coarse debris is entrained during floods and is incorporated into boulder bars (possibly bedrock cored) that are several hundred metres wide and separated by longer, deeper pools (Fig. 8.2b). Many bars do not correspond with the location of tributary junctions and occur at roughly regular intervals (average spacing ~5–7 times gorge floor width) in the 5–6 km reach downstream of the Main Falls. This suggests that bar formation may be controlled by the hydraulics of the occasional high-energy floods in the gorge (cf. Wohl 1992). Farther downstream, bars are less common and less regularly spaced. Beyond 18 km, a distinct gorge is less evident, although the Orange continues to flow in a narrow valley surrounded by deeply dissected terrain.

8.4 Development of the Fluvial Landscape

A possible model for fluvial landscape development can be outlined, which can serve as a hypothesis to be tested in future work.

8.4.1 Waterfall and Gorge Development

The physiography of the Augrabies Falls region (Figs. 8.1c, d, and 8.2b) suggests that the upstream anabranching channel complex originally extended farther west. This is supported by the gravels that can be found in many of the low points in this terrain. Many gravels are of banded iron formation, the nearest outcrops of which are located far to the east, and are commonly subrounded to rounded, suggesting transport during past floods in now-abandoned anabranches. Over time, development of flow concentration along a few dominant anabranches would have led to faster incision, diversion of flow from other anabranches, and further focusing of erosion. Eventually, distinct knickpoints (rapids and/or waterfalls) would have been generated along the dominant anabranches, with upstream knickpoint retreat leading to gorge formation (Fig. 8.3).

In the Augrabies Falls region, two principal knickpoints are now evident, namely the Main Falls and Dry Falls, both of which are located at the head of gorges. The longer gorge downstream of the Main Falls shows that these falls have retreated eastwards faster than the Dry Falls, and the present-day flow distribution indicates that the Main Falls and gorge have captured the bulk of the flow from upstream anabranches. Numerous studies have shown that fluvial bedrock erosion can be driven by a variety of processes, including hydraulic plucking, abrasion, cavitation, and physical and chemical weathering (e.g. Tinkler and Wohl 1998; Whipple et al. 2000). Observations in the Augrabies Falls region show that plucking and abrasion are key contributors to bedrock erosion and knickpoint retreat (Fig. 8.4a–c).

Around the Main Falls, the influence of ongoing knickpoint retreat on low flow patterns in upstream anabranches can be seen clearly (Fig. 8.5). Upstream of these falls, gradient steepening and flow acceleration has led to a faster deepening rate along the primary anabranch, as shown by numerous potholes and other abrasional features developed on elevated bedrock surfaces that are now flooded only irregularly (e.g. Springer et al. 2006). This anabranch has begun to draw flow from its more elevated neighbouring anabranches, thus capturing an increasing proportion of the low flow (Fig. 8.5). The combined flow is then conveyed along a narrow chasm ~10 m deep to the Main Falls, while the downstream reaches of the more elevated anabranches and the Dry Falls typically remain dry, except during the largest floods.

Recognising some of these dynamics, Wellington (1955, p. 360) considered the Dry Falls to be a 'dead channel and

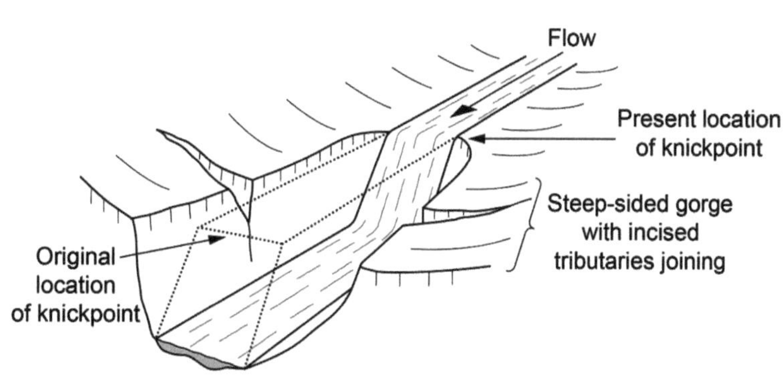

Fig. 8.3 Schematic model showing how upstream knickpoint retreat leads to gorge development (modified after Hayakawa and Matsukura 2003)

Fig. 8.4 Photographs illustrating bedrock erosional processes in the Augrabies Falls region: **a** hydraulic plucking of joint-bounded blocks, evidenced by distinct steps (on extreme *right* of photograph behind the person) and topographic lows (extreme *left*) in the dry channel bed; **b** abrasion in the form of potholes; **c** close-up of exposed bedrock just above the lip of the Main Falls, showing how erosion is accomplished by a combination of plucking and abrasion. Sharp, angular ledges and faces demarcate where blocks have been plucked from the surface (visible in middle of photograph at and above the water surface) while fluting and fine-scale sculpturing of the bedrock (visible in *middle* and on *left* of photograph) and pothole formation (just visible on the *middle* and *upper right* of photograph) provide evidence for abrasion (*Photographs* S. Tooth)

gorge'. Indeed, as the Dry Falls are only active during the largest floods, the Main Falls will continue to erode and thus retreat faster. This will continue a developmental scenario whereby the more elevated anabranches and Dry Falls are gradually abandoned as flow and erosion is concentrated along the anabranches and Main Falls to the south.

8.4.2 Controls on Waterfall Initiation

While the waterfalls in the Augrabies Falls region represent distinctive knickpoints in the long profile of the Orange River (Fig. 8.1b), the controls on their initiation are uncertain as knickpoints can be initiated by various factors (Fig. 8.6). In the Augrabies Falls region, Wellington (1955, 1958) noted how erosion of a capping of quartzite ~40 km north-west of the Main Falls could have initiated knickpoint retreat in the underlying granitoids (cf. Fig. 8.6a). Wellington (1955, p. 360) also acknowledged, however, that the Orange River crosses other resistant lithologies closer to the Main Falls. These could have acted as a temporary barrier, with knickpoint retreat being initiated once the barrier had been breached (cf. Fig. 8.6b). Local tectonic activity and reactivation of faults, shear zones and lineaments across the river course is also a possible explanation for knickpoint initiation (cf. Fig. 8.6c). Alternatively, knickpoints could have been initiated following episodes of continental-scale uplift (Fig. 8.6d) during the Mesozoic or Cenozoic (e.g. King 1951; Wellington 1955; Partridge and Maud 1987; Burke 1996; McCarthy 2013; Kounov et al. 2013). By this explanation, the Main Falls could represent the current position of a knickpoint (or a coalesced series of knickpoints—King 1951, p. 174) that started near the Atlantic Ocean and that has retreated more than 600 km inland over timescales of a few million years to possibly over 100 million years.

Irrespective of the controls on waterfall initiation, knickpoint retreat and gorge formation in an initially low-relief valley floor (Fig. 8.3) can be regarded as representing a renewed stage of landscape denudation. Gorge deepening may initiate a fall in baselevel for tributaries (i.e. anabranches) to the main river, initiating knickpoints in their courses. As these knickpoints retreat along the tributaries, in turn they initiate baselevel falls for their lower-order tributaries, which leads to knickpoint initiation and retreat along their courses, and so on up through the fluvial network.

The physiography west of the Augrabies Falls region provides evidence for long distance knickpoint retreat over at least tens of kilometres. Here, the Orange River flows in a deep valley but a distinct gorge is not evident. Nonetheless, knickpoints (including dry waterfalls) occur along many deeply incised tributaries. For instance, on the Molopo River, which joins the Orange River ~22 km downstream of the Main Falls, a 10–20 m high dry waterfall is located ~12 km upstream from the junction of the two rivers. Landscape denudation is relatively advanced downstream of this waterfall but less advanced upstream (Haughton 1927; Moore 1999). Closer to

Fig. 8.5 Oblique aerial view of the Orange River anabranches in the area above the Main Falls (see area of *whitewater* at lower *left* of image), showing how a small but lower elevation anabranch (*arrow* on the *right*) is drawing water away from larger but more elevated anabranches (*arrows* on the *left*). One of these higher elevation anabranches flows to within 120 m of the edge of the gorge but turns almost 90°(*curved arrow*) to flow back towards the lower elevation anabranch. National Park buildings in the lower right are located about 350 m from the area of whitewater marking the Main Falls (*Photograph* S. Tooth)

Augrabies Falls, many tributaries have been left hanging above the main gorge floor and are only active as waterfalls during large floods, indicating that the knickpoints have retreated only short distances upstream from their junction with the main gorge. Despite local tributary incision and mass failure (Fig. 8.2b), the main gorge sidewalls remain steep and the anabranching complex on the valley floor surrounding the gorge remains largely intact (Fig. 8.1d). Upstream of the Main Falls, a major drop in baselevel has yet to occur, and the valley floor remains relatively undissected, but a future wave of erosion will ultimately drive renewed landscape denudation in this zone. The primary anabranch will continue to outcompete the other anabranches and capture an increasingly larger proportion of flow, and ultimately there will be reversion to a single channel that cascades into a deep gorge.

8.4.3 Timescales and Rates of Waterfall Retreat and Gorge Formation

In the Augrabies Falls region, there are no geochronological data with which to assess the age of waterfall initiation or the rate of waterfall retreat and gorge lengthening. Limited data from other settings worldwide reveal that waterfall retreat rates in relatively weak sedimentary rocks may range from centimetres to a few metres a year (e.g. Hayakawa and Matsukura 2003, 2009), but long-term average retreat rates in resistant igneous or metamorphic rocks tend to be less than a few millimetres a year (e.g. Seidl et al. 1997). Assuming a long-term average retreat rate for the Main Falls of 1 mm/year, the downstream gorge (minimum length of 18 km) would have formed over at least 18 million years. If this average retreat rate was 10 mm/year, then the gorge would have formed over at least 1.8 million years. This retreat would still be imperceptible on a human timescale.

Although timescales and rates of waterfall retreat and gorge formation cannot yet be established, cosmogenic nuclide data have been used to investigate land surface denudation rates in the Augrabies Falls region (Butler 2007). Paired ^{26}Al–^{10}Be analyses show that local denudation rates along anabranches upstream of the falls ($\sim$0.04 mm/year) are faster than on the surrounding interfluves ($\sim$0.003 mm/year) and in the gorge below the Main Falls ($\sim$0.01 mm/year). These denudation rate differences suggest that local

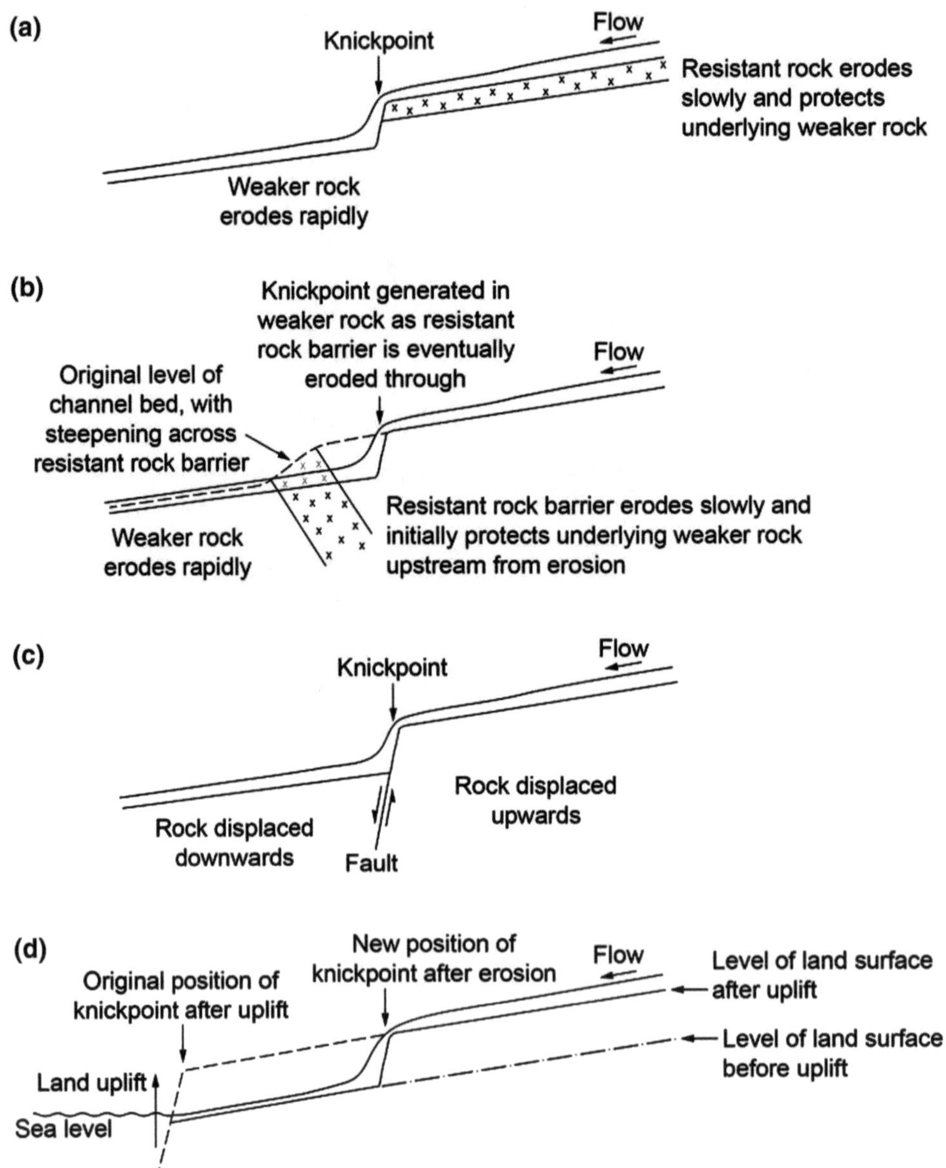

Fig. 8.6 Schematic illustrations of different scenarios by which knickpoints (rapids and waterfalls) can be initiated: **a** and **b** differential erosion of weaker and more resistant rocks; **c** tectonic activity across a river's course; and **d** land uplift. Knickpoints initiated under scenarios a, b and c may be of local to regional significance, whereas those initiated under scenario d may be of continental significance

relief is slowly increasing over time in the vicinity of this major knickpoint, as is suggested by evidence for plucking, potholes and other abrasional features (Fig. 8.4), particularly along the primary anabranch.

8.5 Significance of the Fluvial Landscape in the Augrabies Falls Region

By dint of its large catchment, the long-term behaviour of the Orange River is a key driver in the geomorphological development of southern Africa. Bedrock incision and knickpoint retreat delivers baselevel change into the continental interior, thereby influencing the pattern and tempo of wider landscape denudation. This is exemplified in the Augrabies Falls region, which provides good examples of gorge formation and landscape denudation in the vicinity of retreating knickpoints. Retreat of the Main Falls is extending the gorge and initiating renewed landscape denudation in this part of the valley. While such insights provide opportunities for raising awareness of South Africa's rich geoheritage, with potential benefits in terms of enhancing geoconservation, geotourism and geoscience education, these opportunities have yet to be exploited. At the Augrabies Falls National Park, such insights need to be conveyed to visitors in simplified, widely accessible forms (e.g. through improved signboarding, educational dioramas or computer animations), thereby enhancing the sensory experiences provided by the desert landscape.

Acknowledgments Over the years, collaborative research with various colleagues in the Augrabies Falls region has been supported by the Skye Foundation, De Beers Africa Exploration, the National Science Foundation (USA), and Aberystwyth University. SANParks are thanked for the granting of a research permit to work in the National Park (Project no. 2003-06-18STOO) during the 2003 field season.

References

Bremner JM, Rogers J, Willis JP (1990) Sedimentological aspects of the 1988 Orange River floods. Trans Royal Soc S Afr 47:247–294

Butler JJ (2007) Reconstructing the Quaternary denudation history of the Orange River basin, southern Africa, using cosmogenic ^{10}Be and ^{26}Al analysis. Ph.D. dissertation, University of Edinburgh

Burke K (1996) The African Plate. S Afr J Geol 99:341–409

Council for Geoscience (2013) Earthquake studies in the Augrabies area. Geoclips, December 2013, 7 pp

du Plessis DB, Dunsmore SJ, Burger CE, Randall LA (1989) Documentation of the February–March 1988 floods in the Orange River basin. Technical Report TR 142. Department of Water Affairs, Pretoria

Haughton SH (1927) Notes on the river-system of south-west Gordonia. Trans Royal Soc S Afr 14:225–231

Hayakawa YS, Matsukura Y (2003) Recession rates of waterfalls in Boso Peninsula. Earth Surf Proc Land 28:675–684

Hayakawa YS, Matsukura Y (2009) Factors influencing the recession rate of Niagara Falls since the 19th century. Geomorphology 110:212–216

King LC (1951) South African Scenery: a textbook of geomorphology, 2nd edn. Oliver and Boyd, Edinburgh

Kounov A, Viola G, Dunkl I, Frimmel H (2013) Southern African perspectives on the long-term morpho-tectonic evolution of cratonic interiors. Tectonophysics 601:177–191

McCarthy TS (2013) The Okavango Delta and its place in the geomorphological evolution of southern Africa. S Afr J Geol 116:1–54

Moore AE (1999) A reappraisal of epeirogenic flexure axes in southern Africa. S Afr J Geol 102:363–376

Partridge TC, Maud RR (1987) Geomorphic evolution of southern Africa since the Mesozoic. S Afr J Geol 90:179–208

SANParks (2013) Augrabies Falls National Park. Climate. http://www.sanparks.org/parks/augrabies/tourism/climate.php. Accessed 15 Nov 2013

Seidl MA, Finkel RC, Caffee MW, Hudson GB, Dietrich WE (1997) Cosmogenic isotope analyses applied to river longitudinal profile evolution: problems and interpretations. Earth Surf Proc Land 22:195–209

Slabbert MJ, Malherbe SJ (1983) Die invloed van nate op die loop van die Oranjerivier in die omgewing van die Augrabieswaterval Nasionale Park. Koedoe 26:1–7

Springer GS, Tooth S, Wohl EE (2006) Theoretical modeling of stream potholes based upon empirical observations from the Orange River, Republic of South Africa. Geomorphology 82:160–176

Tinkler KJ, Wohl EE (eds) (1998) Rivers Over Rock: fluvial processes in bedrock channels. Geophysical Monograph Series, vol 107. American Geophysical Union, Washington, DC

Tooth S, McCarthy TS (2004) Anabranching in mixed bedrock-alluvial rivers: the example of the Orange River above Augrabies Falls, Northern Cape Province, South Africa. Geomorphology 57:235–262

van der Walt P (2000) Augrabies Splendour: a guide to the natural history of the Augrabies Falls National Park and the Riemvasmaak wildlife area. Info Naturae, Pretoria

Water Research Commission (1999) Evaporation from the Orange River: quantifying open water resources. Report No. 683/1/99. Water Research Commission, Pretoria

Wellington JH (1955) Southern Africa: a geographical study. physical geography, vol 1. Cambridge University Press, Cambridge

Wellington JH (1958) The evolution of the Orange River: some outstanding problems. S Afr Geogr J 40:3–30

Whipple KX, Hancock GS, Anderson RS (2000) River incision into bedrock: mechanics and relative efficacy of plucking, abrasion and cavitation. Geol Soc Am Bull 112:490–503

Wohl EE (1992) Gradient irregularity in the Herbert Gorge of northeastern Australia. Earth Surf Proc Land 17:69–84

The Richtersveld: An Ancient Rocky Wilderness

9

David L. Reid

Abstract

The Richtersveld, in northwest South Africa, is located at the transition from the coastal plain to the elevated interior plateau and adjacent to the Orange River. Despite this positioning, the primary landscape features of the Richtersveld reflect the much more humid conditions of the Cretaceous period 100 million years ago. Subsequent aridification of the southern African west coast during the Cenozoic resulted in lower landscape denudation rates that left the topography of the Richtersveld largely unaffected, except for periodic changes in climate when increased run-off incised the lower Orange River through the Great Escarpment and into the continental interior. Sea-level fluctuations during the Cenozoic also contributed to river incision, as well as the development of river terraces and marine benches that host economically important diamond placers. Three landscape terrains can be defined along a west–east profile, inland from the coast. The western Richtersveld forms the coastal plain that was cut to near sea level across all lithologies, irrespective of composition and hardness, and is covered by alluvial debris derived from the escarpment and aeolian sands from the coast. The central and eastern terrains form a linear corridor of high topographic relief and dissection that characterises the Great Escarpment. The great antiquity and long-term preservation of the Richtersveld landscape reflects its long geologic and climatic history and is today protected as a wilderness region.

Keywords

Climate change • River incision • Coastal processes • Sea-level change • Cretaceous

9.1 Introduction

In their memoir published on the Richtersveld, De Villiers and Söhnge (1959) described the region as covering some 3,000 square miles ($\sim$7,700 km^2) bounded by the 'big bend' of the lower Orange River, the Atlantic coastline, the 29th parallel and the 17°30′ longitude (Fig. 9.1). Subsequent land use declarations have resulted in the proclamation of the Richtersveld National Park (1991) in the northeast sector, which covers $\sim$1,620 km^2, while a similar area adjacent to the park in the southeast sector, termed the 'Richtersveld Cultural and Botanical Landscape', was inscribed as a UNESCO World Heritage Site (WHS) in 2007 (Fig. 9.1). The Richtersveld is most famous for its high succulent biodiversity and has been named as an Arid Biodiversity Hotspot. Although the region is best known for its rich cultural and biological heritage, the landscape and landforms here are equally impressive with sharp-edged mountains, plains, dry valleys and canyons. The region represents South Africa's only true desert, and yet, this landscape and its landforms have never been previously described in a geomorphological text.

This chapter details (a) the geologic history of the entire Richtersveld region (as defined by De Villiers and Söhnge 1959), which provides a regional context to the smaller

D.L. Reid (✉)
University of Cape Town, Cape Town, South Africa
e-mail: david.reid@uct.ac.za

S. Grab and J. Knight (eds.), *Landscapes and Landforms of South Africa*,
World Geomorphological Landscapes, DOI 10.1007/978-3-319-03560-4_9,

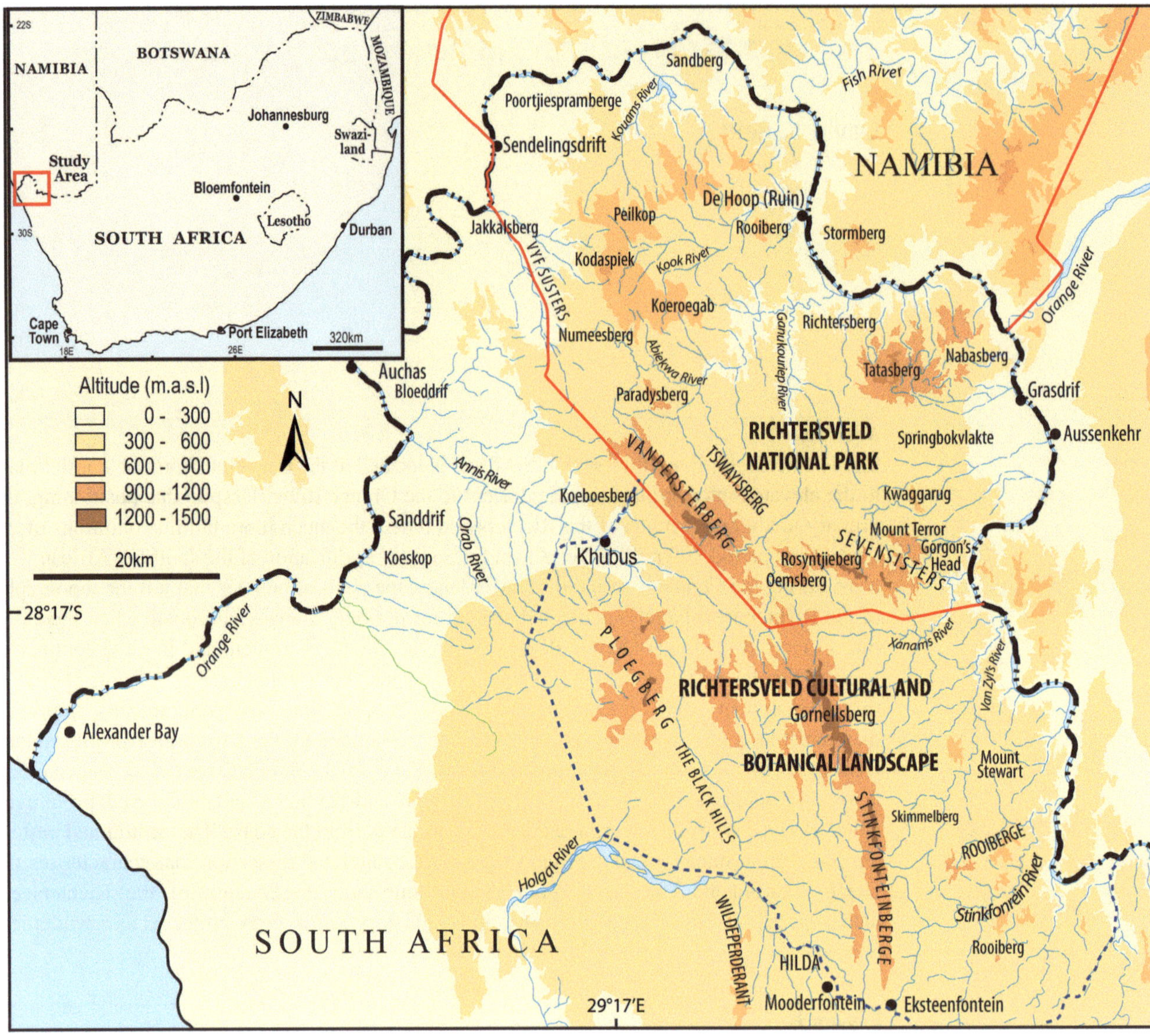

Fig. 9.1 Relief map of the Richtersveld showing the national park and its trans-frontier relationship with Namibia, as well as the World Heritage Site. Various localities are referred to in the text

national park and WHS; (b) the landscape evolution of the region as far as it is currently understood; (c) some of the geologic and climatic controls on the varied landscapes in the region; and (d) the possible role that Richtersveld landscapes may play in offering a possible analogue for planetary geomorphic phenomena and processes.

9.2 Geological History

The Richtersveld is underlain by two major rock sequences, namely the Late Proterozoic Gariep Group and its basement complex (Fig. 9.2). These effectively split the Richtersveld into two halves along a roughly north–south curvilinear boundary, and the constituent bedrock lithologies have exerted a strong control on topography and relief. The pre-Gariep basement complex is a product of at least two major orogenic cycles that culminated in the Rodinia supercontinent, the breakup of which resulted in a rift-drift volcano-sedimentary succession of the Gariep Supergroup (Gresse et al. 2006). Commencing with terrestrial sediments deposited in a rift setting (Stinkfontein Subgroup, ST in Fig. 9.2), the older basement was covered by sandstones and conglomerates that now build the spectacular ridges in the Lekkersing—Eeksteenfontein region of the central Richtersveld. Continued sediment deposition was accompanied by changes in climate and palaeogeography, with glacial conditions prevailing for a period (Kaigas Formation), followed by extensive shallow marine carbonates (Hilda Subgroup, HD in Fig. 9.2). All these contrasting lithologies

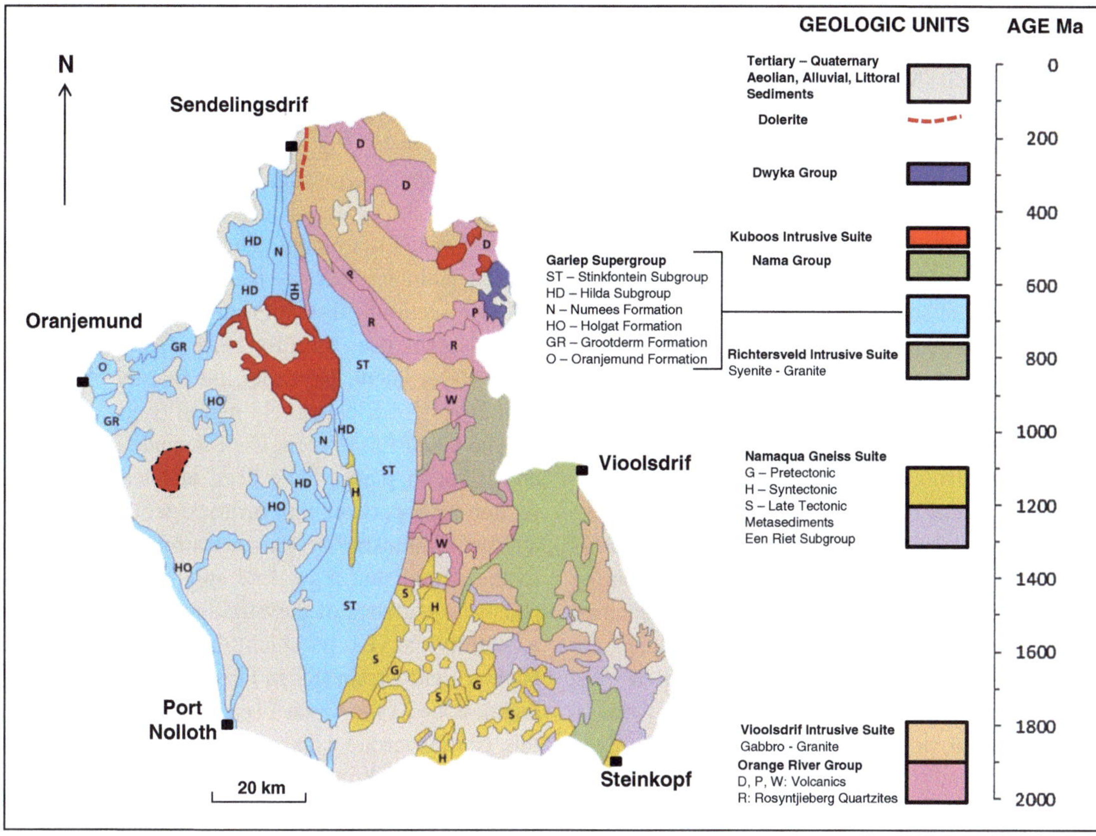

Fig. 9.2 Geological map of the Richtersveld region

(quartzite, conglomerate, limestone) have led to extremely varied landscape responses. A second glacial period around 550 Ma terminated Hilda deposition and mantled the entire landscape of that time with coarse unsorted debris (Numees Formation, N in Fig. 9.2).

The rift and drift episodes were ultimately succeeded by the reversal of plate movements that led to the Pan-African orogenic event, culminating in the Gondwanaland supercontinent. Down-wearing of Pan-African mountain chains led to the deposition in extensive intra-continental foreland basins of the Nama Group (Gresse et al. 2006). At about 500 Ma, an igneous lineament developed in a SW–NE direction across the Richtersveld and resulted in the intrusion of the Kuboos Suite (Fig. 9.2). Erosion of Gondwanaland continued until about 300 Ma with the establishment of a regional low-altitude pre-Karoo peneplain, but with several preserved highland areas (Visser 1989). At this time, much of the supercontinent was positioned at the South Pole, with concomitant glaciation that further promoted regional planation. Ice retreat led to the blanket deposition of the late Carboniferous Dwyka Group tillites and associated shales, which in places were channelled into glacial valleys that would ultimately control the courses of later rivers, such as the Orange (De Wit 1999).

Magmatism associated with continental rifting that led to the opening of the South Atlantic is thought to have taken place over the period 133–128 Ma (Renne et al. 1996), and while the majority of activity is confined to the northwest part of Namibia, some intrusive activity extended down to the Richtersveld (Reid and Rex 1994), in the form of dolerite dykes near Sendelingsdrif (Fig. 9.2). For the past 120 Ma, since the establishment of the South Atlantic Ocean, the Richtersveld has occupied a near coastal setting, with limited progradation despite the presence of a major delta receiving sediment from South Africa's largest river system.

9.3 Long-term Landscape Evolution

Southern Africa appears to have occupied an elevated position within the Gondwanaland supercontinent during Triassic and Jurassic times, partly due to orogenic activity associated with the Cape Fold Belt, which was subsequently enhanced by upwelling mantle plumes and convection cells that led to continental break-up. The Richtersveld was probably located at the margins of plume-induced lithospheric uplift that occurred concurrently with the activity of the Karoo Plume at ~180 Ma and the Etendeka–Parana

Plume at ~130 Ma, but its proximity to the Atlantic rifted margin probably ensured significant elevation during the early Cretaceous as well (McCarthy 2013). Prior to plume-induced domal uplift and rift margin uplift, the Richtersveld was most likely covered with several km of Karoo strata and flood basalts at the time of Cretaceous rifting. Thermally induced domical uplifts developed above the Karoo and Parana plumes controlled Jurassic–Cretaceous drainage systems that led to the asymmetric drainage and preferential erosion of the interior of southern Africa (McCarthy 2013). This was superimposed on the rifted margin profile that developed along the Atlantic coast (Kuonov et al. 2009).

A period of accelerated denudation of the southern African subcontinent during the middle–late Cretaceous was enhanced by increased run-off due to the warm humid climate that persisted at that time. A coastal plain would have developed along the Atlantic rifted margin, with development of the Great Escarpment initially by back-wearing (traditional scarp retreat), but due to major rivers breaching this topographic barrier and eroding the interior plateau, the inland migration of the escarpment was constrained by a drainage divide (Decker 2010), leading to control by down-wearing from opposing sides of the plateau (van der Beek et al. 2002). Consequently, the escarpment developed into a broad belt of high relief and dissection, parallel to the coast but punctuated by embayments that represent preferential erosion by major rivers from the interior, perhaps influenced by pre-Karoo glaciated topography as the Karoo strata were stripped off the basement. The Richtersveld occupies a unique section of the Great Escarpment, given that it coincides with the Orange River's entry through this topographic barrier and later exit into the Atlantic at Oranjemund. The 'big bend' followed by the lowermost stretch of this river has caused the escarpment to be excavated from both river banks (see profiles in Figs. 9.3 and 9.4).

Denudational rates decreased after the Late Cretaceous, as evidenced in the offshore sedimentary record, where thick Cretaceous sequences are succeeded by a thinner Tertiary record, but there is debate concerning the cause, with reduced precipitation and run-off being regarded as more important than lowered river course elevation (Decker et al. 2011). The onset of drier arid conditions slowed denudation considerably to a point where little significant change to the Cretaceous landscape (both in terms of elevation and relief) occurred throughout the entire Cenozoic, except for brief periods of warmer humid conditions that could have promoted renewed river incision. Smaller scale epeirogenic and far-field extensional tectonics may have occurred during the Cenozoic and may still be active today, but only in the flat continental interior could small vertical movements have a landscape impact, in the form of reorganizing river drainage systems (McCarthy 2013). Such reorganization could increase run-off and incision without the need for changes in climate.

9.4 Landscape Terrain Classification

Elevation profiles have been compiled in east–west and north–south profiles in order to illustrate the varied terrain of the Richtersveld (Fig. 9.3). Sections C1–C8 are juxtaposed to display the relationship between the current coastline, the position of the Orange River valley and the dissected nature of the Great Escarpment. The subdued slope profile of the coastal plain is clear, as is its interruption by the Orange River in sections C1–C4. From Oranjemund southwards, the coastal plain rises steadily before meeting the foot of the Great Escarpment. Hereafter, the coastal plain is referred to as the western Richtersveld region (Fig. 9.4).

The Great Escarpment itself is made up of high-elevation areas that in places have been affected by deep river dissection. Comparison with bedrock geology (Fig. 9.2) reveals a close relationship between escarpment relief and lithology to the extent that a twofold subdivision is possible. Hard resistant quartzites of the Stinkfontein Subgroup crop out in a slightly curved mountain belt that defines the central Richtersveld terrain (ST in Fig. 9.2) and thus forms the west-facing edge of the Great Escarpment. The eastern margin of this central terrain is a major geological terrain boundary, where the Stinkfontein quartzites give way to the older basement complex which underlies the eastern Richtersveld landscape terrain. Cutting through all these Richtersveld landscape terrains is the incised canyon of the lower Orange River (Fig. 9.5), with its tributaries laterally dissecting the Great Escarpment and continuing into the continental interior.

9.4.1 Orange River Canyon

The 'big bend' in the Orange River was probably controlled by the presence of a basement horst that deflected the ancestral course of the river after the softer overlying Karoo strata were removed. The magnitude of the deflection may have been tempered by crossing faults in the north-trending basement horst, which allowed the river to return southwards and exit into the Atlantic at Oranjemund. Deflection of the river course might also have occurred as the glacially channelized pre-Dwyka surface was re-exhumed, prior to exposure of the basement. Superimposed on the 'big bend' are many smaller scale meanders which must have been inherited from the subdued landscape that existed when the ancestral Orange River was flowing over Karoo strata. Several generations of diamondiferous river terraces containing Miocene–Pliocene fossils occur along the lower Orange River. It is likely that increased run-off during the Tertiary (and possibly extending into the Pleistocene) featured prominently in the transport of diamonds from the denuding continental interior.

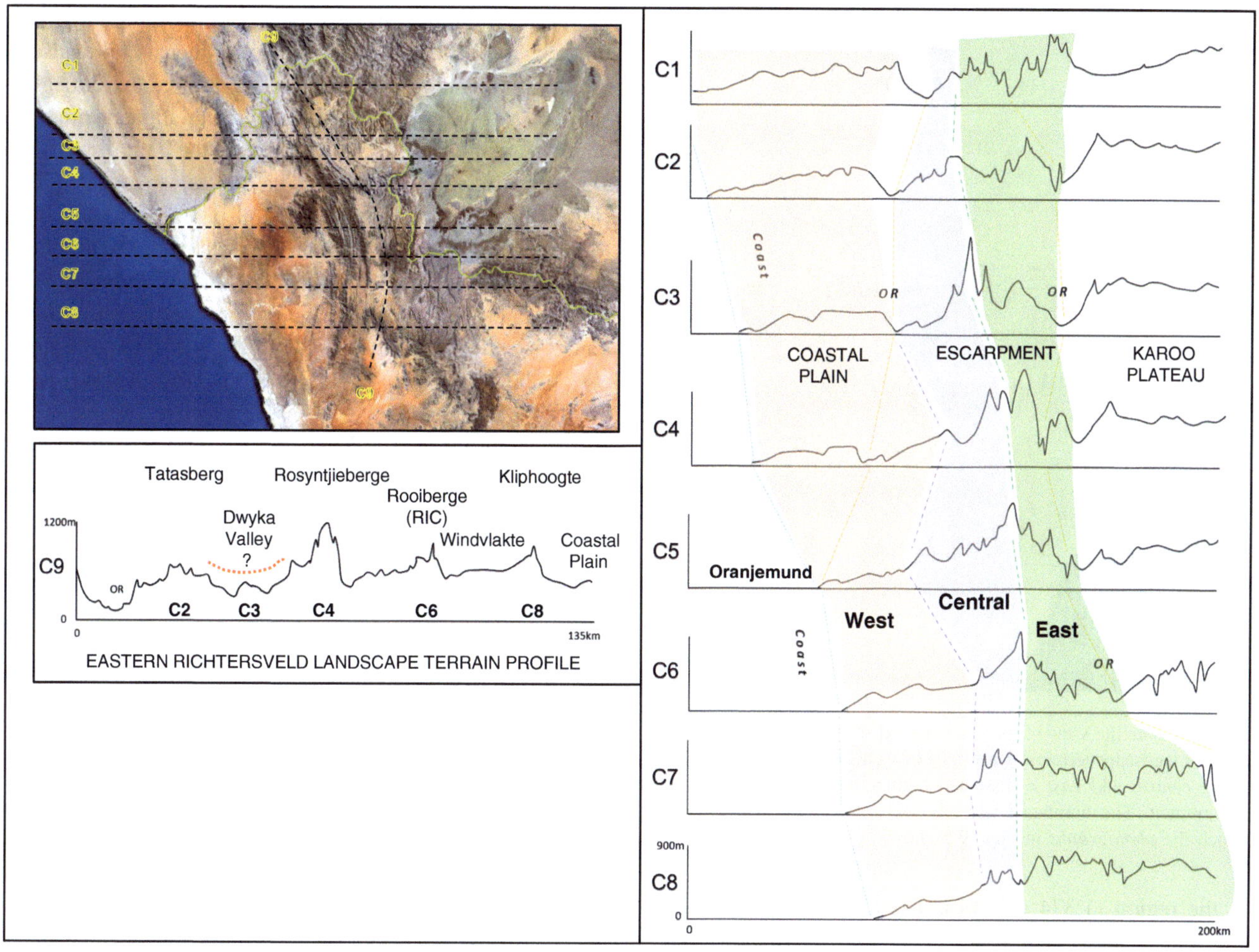

Fig. 9.3 Topographic relief profiles across the Richtersveld. Superimposed on the E–W line profiles (*C1–C8*) is the present-day coastline, and the trace of the Orange River (OR) and the boundaries separating the three landscape terrains identified: west (=coastal plain); central and east (=Great Escarpment). Profile *C9* shows the N–S topographic variation within the east terrain, the strong influence of lithology and incision of the Orange River

9.4.2 Eastern Richtersveld

As one proceeds from north to south, the eastern Richtersveld commences as a high-altitude mountainous domain, representing the most complex of the three landscape terrains, with down-wearing exploiting the highly variable lithologies (Fig. 9.3, profile C9). Farther south, the lower elevation and topographically subdued Springbokvlakte contrast with its surrounding wide, sandy valleys and low granite hills (Fig. 9.6). It is probable that part of the lower elevation has been inherited from a pre-Karoo landscape carved by Dwyka glaciers (Fig. 9.3, profile C9). The southern boundary of the Springbokvlakte is an abrupt transition to a narrow but precipitous belt of quartzite ridges that mark the Rosyntjieberge (Fig. 9.6), which crosses the Orange River at its narrowest point in the canyon and probably represents the most spectacular stretch of this watercourse. South of the Rosyntjieberge is the composite landscape domain of the Windvlakte (Fig. 9.7). Dissection of the Windvlakte plateau by two rejuvenated tributaries, the Ga-ams and Stinkfontein rivers, has caused its distinctive profile (Fig. 9.3, profile C9).

At the southern boundary of the Richtersveld, the Great Escarpment turns inland and terminates at Kliphoogte (Fig. 9.3, section C9), representing an expansion of the coastal plain. At this locality, the distinction between the central and eastern Richtersveld landscape terrains disappears and lithological control becomes less apparent. From Port Nolloth inland, the subdued coastal plain extends ~70 km to the foot of the Anenous Pass.

9.4.3 Central Richtersveld

The central landscape terrain is the mountainous spine of the Richtersveld and contains Cornellsberg, the highest summit

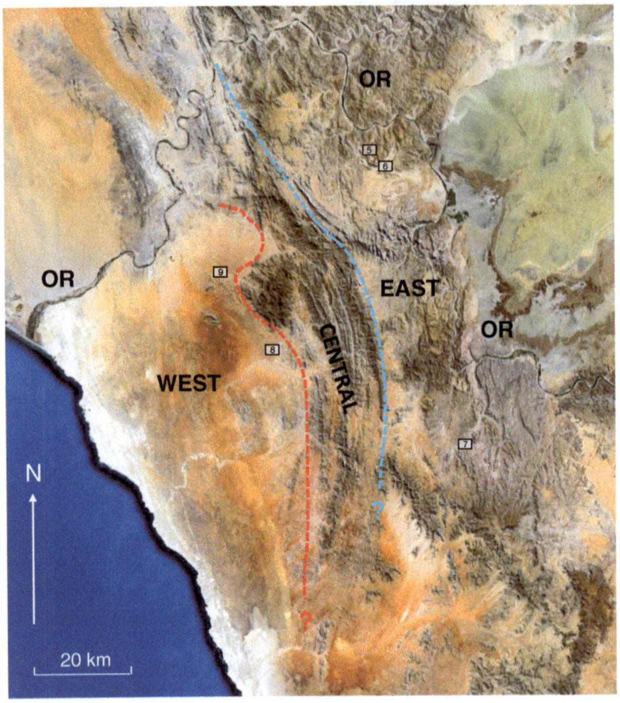

Fig. 9.4 Satellite view of the Richtersveld showing the distribution of the four landscape terrains identified: *OR* = Orange River; *west* = coastal plain; *central* and *east* = dissected high relief corridor of the Great Escarpment. The numbered sites (boxes) refer to the positions from which the *photographs* in Figs. 9.5, 9.6, 9.7, 9.8 and 9.9 were taken

in the region (1,374 m a.s.l.). The terrain forms a slightly curvilinear arc trending north to north-northeast and is underlain by resistant quartzites of the Stinkfontein Subgroup (Fig. 9.7).

The western margin of the central terrain is also lithologically controlled. Here, the Stinkfontein quartzites are succeeded by softer Hilda carbonates that have developed karstic and highly ribbed surface features through dissolution (Fig. 9.8). In the northern sector of the central landscape terrain, the Kuboos granite massif builds the Ploegberg summit and interrupts the linear strike-aligned ridges of quartzite, shale and limestone. In plan, the Kuboos intrusive element is nearly circular and has a diameter of ~30 km (Fig. 9.1), of which the western half has been planed down to the level of the coastal plain and is covered with alluvial and aeolian sands (Fig. 9.2). West of the Ploegberg, the coastal plain covers another young granite pluton at Swartbank, which is described as being very similar to Kuboos (De Villiers and Söhnge, 1959), so its erosion down to the level of the plain can be taken as evidence for lateral planation from the coast. More recent sea-level fluctuations have been responsible for grade changes and dissection of the Annisvlakte fan slopes (Fig. 9.9).

9.4.4 Western Richtersveld

The western terrain region covers the coastal plain that is bordered by the current coastline and the arcuate western boundary of the central terrain highlands. It narrows to the south as the Stinkfontein quartzite ridges approach and reach the coast at Kleinzee. To the north, it crosses the lowermost stretch of the Orange River, forming the Sperrgebiet (German for 'Prohibited area'—diamond mining area), and ultimately merges with the Namib Sand Sea. The subdued landscape of the western terrain is considered to have not changed significantly since the Cretaceous, when it was formed by denudation below the Great Escarpment. The later stages of its history (from the Early Tertiary) involved minor landscape responses to periodic changes in sea level and increased runoff. Most important has been the distribution and deposition of

Fig. 9.5 View from near the summit of Tatasberg, looking east towards the Orange River at Aussenkehr (R) and the downfaulted Karoo cover sequence (K) intruded by a Jurassic dolerite sill (D) (*Photograph* D. Reid)

Fig. 9.6 View south from the slopes of Tatasberg across the Springbokvlakte to the imposing wall of vertical quartzite strata of the Rosyntjieberge (*Photograph* D. Reid)

Fig. 9.7 View west from the eastern terrain (Windvlakte) across to the Stinkfonteinberge, the eastern margin of the central terrain (*Photograph* D. Reid)

alluvial gravel terraces along the course of the lowermost Orange River, and coastal wave and current action. High-level strandlines developed during transgressions and concentrated gem-quality diamonds into placers that were redistributed along the coast after being transported from the interior by the Orange River. Those deposits, located south of the current Orange River mouth, are thought to have been transported northwards by longshore currents from river sources to the south that drained the continental interior, such as the Buffels and ancestral Karoo River (De Wit 1999). The present mouth of the Orange River is thought to have attained its present position during the Palaeocene–Eocene, after which another diamond placer started to develop from Oranjemund northwards along the Namibian coast.

Fig. 9.8 View SE from the coastal plain (western Richtersveld terrain) towards subdued ridge and valley outcrops of the Hilda carbonates that mark the western margin of the central terrain (*Photograph* D. Reid)

9.5 The Richtersveld as a Planetary Landscape

Tourist literature often describes remote, arid, largely uninhabited, terrestrial desert landscapes as 'lunar, moon-like or Martian', which clearly alludes to their visual geomorphic properties. Indeed, many Moon and Mars exploratory missions have involved familiarisation with such desert environments, in order to prepare astronauts, test survey equipment and procedures, and assist in the interpretation of results transmitted back to Earth.

Our accumulated knowledge of the Moon has revealed that its surface is arid, mantled by widespread and often thick deposits of impact and volcanic ejecta (lunar 'soil'), with solid bedrock confined to the lunar *mare* (basalt lava) and highlands (anorthosite crust). Probably the only aspect that is common to both the lunar landscape and that of the Richtersveld is their largely relict nature, in the sense that the latter preserves a terrestrial façade generated in the Cretaceous (90–60 Ma ago) and has experienced limited surficial modification since that time. Apart from a few relatively small impacts and *mare* eruptions, the lunar landscape has likewise been preserved for $\sim$3 Ga (Hörz et al. 1991).

Mars, however, is turning out to possess landscape elements that may show closer similarities with that of the Richtersveld. There is now evidence for ancient waterlain sediments and thus confirms interactions between the Martian lithosphere, hydrosphere and atmosphere (Chapman 2011). Ancient relict landscapes originally carved by liquid water have been recognised and subsequently partially modified by impact cratering, aeolian processes and slope failure. Planetary geologists are increasingly interested in landscapes and their evolution (especially of Mars), and it would be advantageous if such scientists considered terrestrial landscapes such as the Richtersveld, as well as in broadening our understanding of hyper-arid mountain geomorphic dynamics. In particular, there is a need to determine micro-scale weathering (products of which are visible in Fig. 9.5), erosion and depositional processes currently operative in the region.

Fig. 9.9 View NE across broad alluvial fans (Annisvlakte) emanating from the Kuboos granite (to the right of the photograph), showing dissection due to fluctuating sea levels which move the baseline of the western landscape terrain (*Photograph* D. Reid)

Acknowledgments This chapter was improved considerably through the efforts of Mike De Wit, Stefan Grab, Jasper Knight and one anonymous reviewer. Research in the Richtersveld since 1972 has been supported by the South African NRF, its predecessor FRD, as well as UCT. This chapter is dedicated to the late Professor John De Villiers, who first introduced me to one of his most favourite places on Earth.

References

Chapman M (ed) (2011) The geology of Mars, evidence from earth-based analogs. Cambridge University Press, Cambridge

De Villiers J, Söhnge PG (1959) The geology of the Richtersveld. Memoir Geological Survey South Africa 48, 295 pp

De Wit MC (1999) Post-Gondwana drainage and the development of diamond placers in western South Africa. Econ Geol 94:721–740

Decker JE (2010) Landscape evolution and equilibrium in southern Africa: insights from cosmogenic noble gases in Karoo dolerites and geospatial analyses. Unpublished PhD thesis, University of Cape Town

Decker JE, Niedermann S, De Wit MJ (2011) Soil erosion rates in South Africa compared with cosmogenic ^{3}He-based rate of soil production. S Afr J Geol 114:475–488

Hörz F, Grieve R, Heiken G, Spudis PD, Binder A (1991) Lunar surface processes. In: Heiken G, Vaniman D, French B (eds) Lunar sourcebook. Cambridge University Press, Cambridge, pp 61–120

Gresse P, Von Veh M, Frimmel H (2006) Namibian (Neoproterozoic) to early cambrian successions. In: Johnson M, Anhaeusser C, Thomas R (eds) The geology of South Africa. Geological Society South Africa/Council for Geoscience, Johannesburg/Pretoria, pp 395–420

Kuonov A, Viola G, De Wit MJ, Andreoli MA (2009) Denudation along the Atlantic passive margin: new insights from apatite fission-track analysis on the western coast of South Africa. In: Lisker F, Ventura B, Glasmacher U (eds) Thermochronological methods: from palaeotemperature constraints to landscape evolution models. Geol Soc London (Special Publication) 324: 287–306

McCarthy T (2013) The Okavango delta and its place in the geomorphological evolution of southern Africa. S Afr J Geol 116:1–54

Reid D, Rex D (1994) Cretaceous dykes associated with the opening of the South Atlantic: the Mehlberg dyke, Northern Richtersveld. S Afr J Geol 97:135–145

Renne P, Glen J, Milner S, Duncan A (1996) Age of Etendeka flood volcanism and associated intrusions in southwestern Africa. Geology 24:659–662

Van der Beek P, Summerfield M, Braun J, Brown R, Fleming A (2002) Modeling post breakup landscape development and denudational history across the southeast African (Drakensberg Escarpment) margin. J Geophys Res B12: ETG 11-1—ETG, pp 11–18

Visser J (1989) The Permo-Carboniferous Dwyka Formation of southern Africa: deposition by a predominantly subpolar marine ice sheet. Palaeogeogr Palaeoclimatol Palaeoecol 70:377–391

The Cederberg: A Rugged Sandstone Topography

10

Lynne J. Quick and Frank D. Eckardt

Abstract

The Cederberg forms part of the western branch of the Cape Fold Belt, a mountain range that resulted from orogenic (mountain-building) processes in the Permo–Triassic ($\sim$300–230 Ma ago). After deposition, the Ordovician to Carboniferous sandstones and shales of the Cape Supergroup were subjected to faulting, folding and subsequent weathering which has produced a rugged mountainous terrain characterised by a sequence of elevated ridges and peaks (up to 2,027 m a.s.l.) separated by broad linear valleys. The geomorphology of the region is strongly controlled by these bedrock structures, which illustrates the close relationship between geologic and geomorphic patterns of landscape evolution over long timescales. The topography of the region has also exerted control on the Cederberg's Mediterranean climate, with winter rains that support the Fynbos and Succulent Karoo biomes. The interlinked geology, geomorphology and ecology are protected as part of the Cederberg Wilderness Area, which is a significant geotourism and geoheritage region, rich in archaeological remains.

Keywords

Fynbos • Cape Fold Belt • Table Mountain Group • Sandstone weathering

10.1 Introduction

The geomorphic evolution of the Cederberg is related to the formation of the Cape Fold Belt (CFB), of which it forms the north-western arm, and was shaped by major folding and faulting events associated with the Cape Orogeny, a mountain-building period in the Permian and Triassic ($\sim$300–230 Ma). The Cederberg is well known for its spectacular rock formations, a result of weathering and erosion of the folded and faulted Ordovician to early Devonian sandstones and mudstones of the Table Mountain Group (TMG). The geomorphology of the region, by contrast, is very poorly known, although its climate history during the Pleistocene and Holocene has been reconstructed using a range of biological records (e.g. Quick et al. 2011; Valsecchi et al. 2013).

The region has been subject to human occupation for millennia and has well-preserved examples of San rock art in caves and overhangs (Orton and Mackay 2008). Originally, the area was referred to as the 'Cedarberg' in reference to the endemic, and now critically endangered, Clanwilliam Cedar (*Widdringtonia cedarbergensis*) (Mustart et al. 1995). The currently accepted 'Cederberg' is a combination of the English and Afrikaans ('Sederberg') spellings, while the geologic formation that forms the Cederberg's distinctive shale band remains the Cedarberg Formation.

10.2 Environmental Setting

The Cederberg forms part of the drier north-western limb of the CFB of the south-western Cape, South Africa (Fig. 10.1). This mountainous region lies generally between 1,200 and

L.J. Quick (✉) · F.D. Eckardt
University of Cape Town, Cape Town, South Africa
e-mail: lynne.j.quick@gmail.com

S. Grab and J. Knight (eds.), *Landscapes and Landforms of South Africa*,
World Geomorphological Landscapes, DOI 10.1007/978-3-319-03560-4_10,

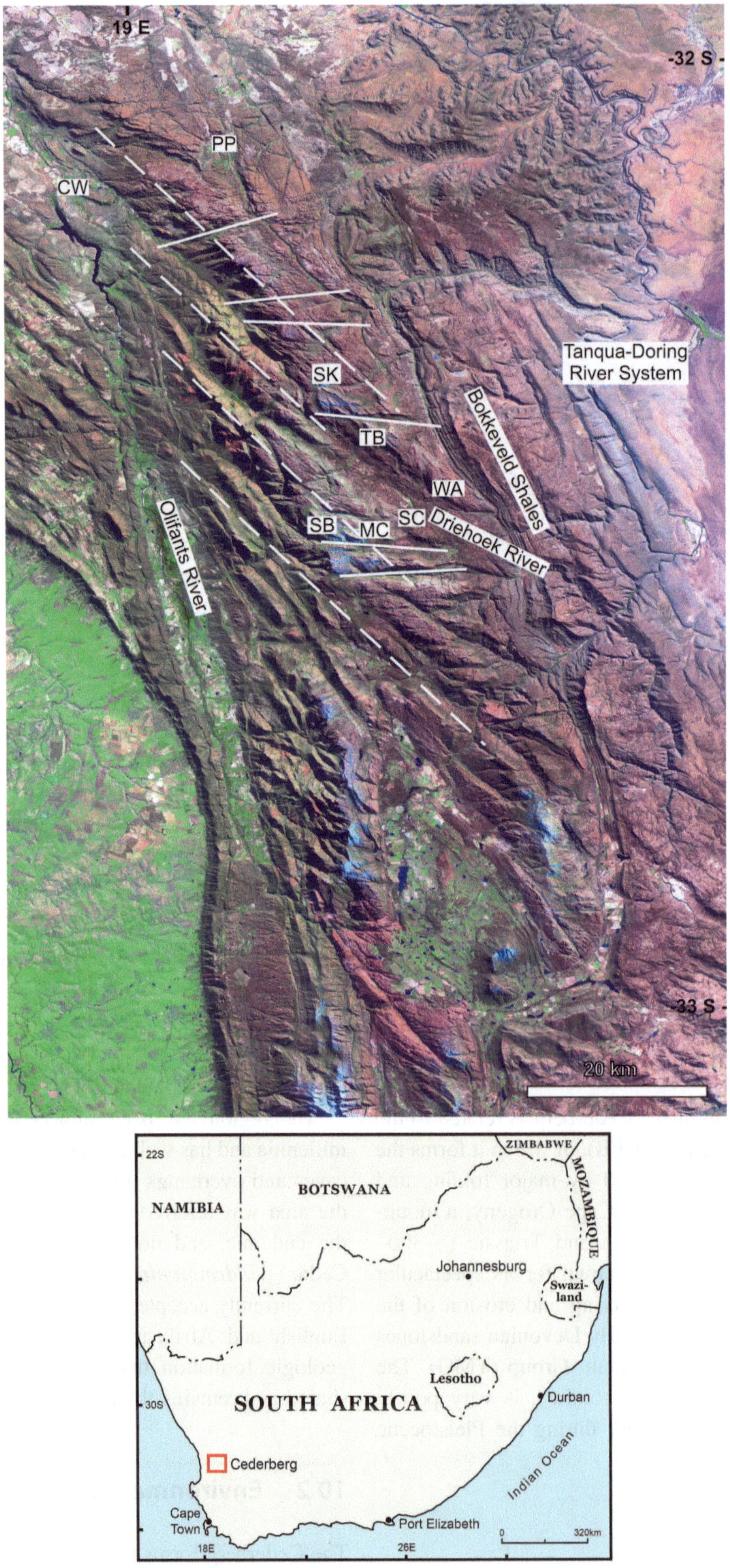

Fig. 10.1 Landsat 742 false colour image of the Cederberg anticline, located between the Olifants River valley to the west and the Bokkeveld Group shales of the Tanqua Karoo to the east. Major peaks include Sneeukop (SK, 1,930 m), Tafleberg (TB, 1,969 m) and Sneeuberg (SB, 2,027 m). The sedimentary rocks of the Table Mountain Group were deformed into upright folds and were dissected subsequently by northwest-striking faults (*dashed lines*) and several east–west-oriented cross-faults dissecting it (*solid lines*). *PP* Pakhuis Pass, *CW* Clanwilliam, *WA* Wolfberg Arch, *SC* Stadsaal Caves and *MC* Maltese Cross (image from Global Landcover Facility [GLCF])

1,500 m a.s.l. and is situated approximately 250 km north of Cape Town. The Cederberg is delimited by two river catchments: to the west is the Olifants River and to the east the Tanqua–Doring river system. The Driehoek River is the major river system within the central Cederberg, flowing through the Driehoek valley before turning eastwards and entering the Doring River (Fig. 10.1).

Located within southern Africa's winter rainfall zone, the Cederberg presently experiences a Mediterranean type climate with warm, dry summers and cold wet winters. During winter, temperatures often drop below freezing and the mountain peaks are frequently covered in snow. The topographic complexity of the region (high ridges and peaks, broad linear valleys) leads to considerable spatial and temporal rainfall variability. For example, Pakhuis Pass, towards the north-eastern boundary of the Cederberg receives an average of 250 mm per year, the central Cederberg (Driehoek valley) is associated with an average of 500 mm per year, and the higher peaks may receive over 1,000 mm per year (Schulze 1997; Hijmans et al. 2005).

The underlying geology and soil characteristics of the Cederberg (together with the other mountain ranges that constitute the CFB) is a major determining factor for the distribution of Mountain fynbos, a highly diverse vegetation type (e.g. Campbell and Werger 1988; Cowling and Lombard 2002). The Cederberg also marks the boundary (ecotone) between the fynbos and Succulent Karoo biomes. A large portion of the Cederberg has been proclaimed the Cederberg Wilderness Area and obtained World Heritage status in 2004 as part of the wider Cape Floral Region Protected Areas (CapeNature 2013).

10.3 Geology

The siliciclastic sedimentary rocks of the Cape Supergroup were deposited on a passive continental margin in a variety of terrestrial (fluvial) and shallow marine depositional environments during the Early to Middle Palaeozoic (Shone and Booth 2005). The Cape Supergroup comprises a succession of sandstones, shales and minor conglomerates subdivided into the Table Mountain, Bokkeveld and Witteberg Groups (Broquet 1992). The oldest and texturally/mineralogically mature sandstone-dominated TMG lies unconformably on Precambrian–Cambrian basement rocks, such as the Malmesbury Group shales and Cape Granites (e.g. du Toit 1954; Visser 1974). The Cederberg is composed predominantly of rocks belonging to the TMG. To the west, near the Olifants River, there are remnants of the older Malmesbury Group shales and outliers of the younger Bokkeveld Group shales. To the east, there is a sharp transition from the sandstone-dominated formations of the TMG, to the shale-dominated formations of the Bokkeveld Group along the Moordenaarsgat and Doring rivers. Here, the Table Mountain sandstones dip steeply beneath the shales of the Bokkeveld Group (Fig. 10.2a). The Cederberg is dominated by three major formations of the TMG, namely the Peninsula, Pakhuis and Cederberg Formations and the Nardouw Subgroup (Table 10.1).

The Peninsula Formation is the thickest unit (up to 3,500 m thick) in the TMG, comprising of quartzitic sandstones with minor siltstone and conglomerate (Visser 1974; Broquet 1992; Gresse and Theron 1992). This formation is found throughout the Cederberg, forming the thick, basal sandstone unit in the region (Fig. 10.2b, c). Overlying the Peninsula Formation is the much thinner (~40–70 m) Pakhuis Formation (Gresse and Theron 1992) which contains evidence for glacial processes at the time of its deposition during the late Ordovician (Fig. 10.3). It comprises a green-blue or reddish mudstone without lamination, but contains randomly scattered pebbles and boulders with flattened and striated surfaces (du Toit 1926; Rust 1967). There is some debate surrounding the deposition and deformation of the Pakhuis Formation (and the fold zone), calling into question the earlier theories relating to the direct influence of local ice-sheet movement (e.g. Rust 1967; Sutcliffe et al. 2000; Rowe and Backerberg 2011).

Gradationally overlying the Pakhuis diamictites is the Cedarberg Formation, a dark, fine-grained, richly fossiliferous shale (Soom Member) succeeded by slightly coarser grained mudstones and siltstones (Disa Member) (Broquet 1992). The relatively thin Cedarberg Formation ranges from ~50 to 120 m in thickness (Broquet 1992) and is locally known as 'Die Trap' (the step), because it forms a regional marker with a distinct vegetation visible as a narrow green band between the more sparsely vegetated blocky sandstones of the Peninsula Formation and the Nardouw Subgroup (Fig. 10.2b, c). The exceptionally well-preserved fossils, including remarkable soft-tissue forms (e.g. Aldridge et al. 2001), point towards a marine environment during the deposition of the postglacial Cedarberg Formation.

The overlying Silurian Nardouw Subgroup is dominated by cross-bedded quartz arenites in tabular, sheet-like layers that are distinctly redder than the sandstones of the older Peninsula Formation (Fig. 10.2b, c). It is thought that sandstones of the Peninsula Formation and Nardouw Subgroup resulted from deposition in high energy, fluvial depositional systems that were subjected to periodic marine inundation and reworking (Broquet 1992; Tankard et al. 2009). Most of the conspicuous peaks in the Cederberg comprise quartz arenites of the Nardouw Subgroup which has a cumulative thickness of 700 m.

Fig. 10.2 Cederberg landscapes: **a** the contact with the Bokkeveld shales at the eastern boundary of the Cederberg (*Photograph* L. Quick), **b** view of Tafelberg with the Driehoek River valley in the foreground (*Photograph* D. Bonora and T. Hoffman), **c** Sneeuberg, the highest mountain peak (2,027 m) in the Cederberg (*Photograph* L. Quick) and **d** Uitkyk Pass, an example of a characteristic Cederberg broad river valley surrounded by rugged peaks (*Photograph* D. Bonora and T. Hoffman)

Fig. 10.2 (continued)

Table 10.1 Major geological units found in the Cederberg from oldest (bottom) to youngest (top)

Geologic formation	Description	Location
Nardouw Subgroup	Coarse-grained orthoquartzites with some pebbles and lenses of vein quartz, distinctively redder than the Peninsula Formation	Plateau-like summits above the shale band, e.g. Tafelberg, Sneeukop and Pakhuis Peak
Cederberg Formation	Shale and siltstone interbedded with fine-grained sandstone can be identified in the landscape as a narrow grass-covered band strongly contrasted by the rocky sandstone surfaces above and below it	Band running below all major peaks
Pakhuis Formation	Thin layer of diamictite or glacigenic mudstones found immediately below the shale band, containing varvites and argillaceous dropstones which show polish and striae	Remnants of the glacial pavement, examples of which can be found along the road that passes over the Pakhuis Pass, on the mountain slopes at Klein Vlei and Groenberg
Peninsula Formation	Thick deposit of light-grey, coarse-grained quartzitic sandstones interspersed with occasional white quartz pebbles and sand-shale lenses	Characterises all regions of the Cederberg where the upper strata have been eroded away

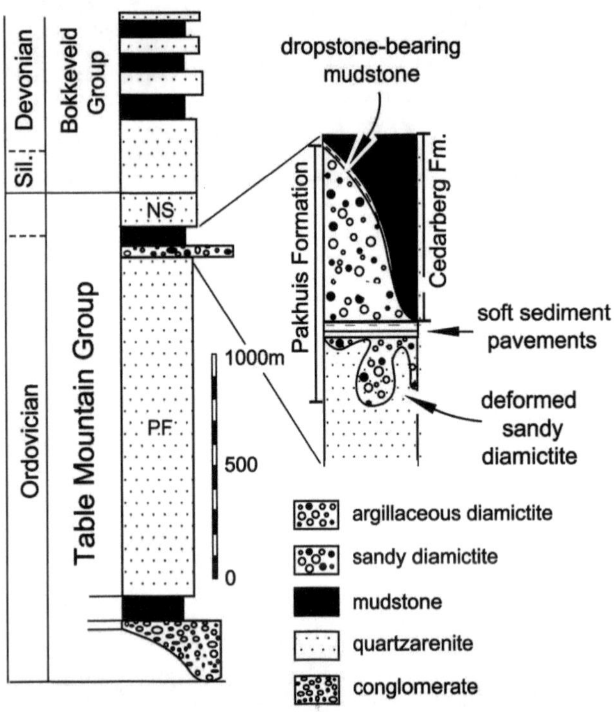

Fig. 10.3 The contact between the Pakhuis and the Cedarberg Formations (adapted from Young et al. 2004). *NS* Nardouw Subgroup and *PF* Peninsula Formation

10.4 Geologic Controls on Cederberg Topography

There is a close relationship between patterns of bedrock geology, tectonics and present-day topography and landscape. Following deposition of the Cape Supergroup from ~300 to 230 Ma, mountain-building (tectonic) forces deformed the southern African region, giving rise to the CFB (King 1963; de Wit and Ransome 1992; Shone and Booth 2005). The CFB can be partitioned into three main domains based on the types of folding, structural grain and amount of shortening (de Beer 1995). The north-trending western branch (which includes the Cederberg) is characterised by relatively open, upright, first-order folds and monoclines. In contrast, the east-trending southern branch is much more deformed with north-verging, recumbent first-order folds and a high incidence of second-order folds and the Cape Fold Syntaxis, the area where the two branches intersect (de Beer 1992, 1995). All three domains are further altered by post-tectonic normal faults that are primarily associated with the break-up of the supercontinent Gondwana (de Wit and Ransome 1992; Shone and Booth 2005).

The Cederberg forms part of the western branch of the CBF, representing a fairly isolated broad anticlinal ridge. The TMG sediments, of which the area is comprised, were deformed into open, upright folds and overprinted by northwest-striking basement-involved faults (see Tankard et al. 2009), with many west-southwest-oriented cross-faults dissecting its western margin (Fig. 10.1). The major topographic features of the Cederberg therefore also trend north–south. The quartzitic sandstones of the Peninsula Formation and Nardouw Subgroup are relatively resistant to weathering, whereas the Cedarberg shales and mudstones are significantly less resistant. The differential resistance and weathering/erosion of these geological formations has resulted in a rugged topography characterised by a complex sequence of elevated ridges and peaks, separated by broad linear valleys with the widely prevalent characteristic shelf or step-like patterns of the shale band (Figs. 10.1 and 10.2b–d).

The more resistant TMG sandstones are structurally less deformed; thus, upland summits developed here are often nearly flat (e.g. Tafelberg: Fig. 10.2b).

10.5 Large-Scale Geomorphology

Landscapes of the Cederberg, corresponding to large-scale patterns of weathering and erosion, reflect underlying geologic patterns quite closely. For example, weathering of softer shales has resulted in more subdued topography and deeper, richer soils, whereas weathering of sandstones has resulted in thinner and better drained acidic soils which have low nutrient levels. Across the Cederberg, river patterns and slope processes strongly reflect geologic control. Throughout, the orientation of river valleys often coincides with the alignment of faults (Fig. 10.1). In upland areas, rivers are usually located in V-shaped valleys and commonly meet larger rivers at 90° angles. These upland rivers commonly flow across bedrock strike, which suggests they are superimposed features set down upon underlying rock structures as these were exposed by the erosion of overlying rocks. This includes the Tanqua–Doring river system (Fig. 10.1), which was also affected by river capture as the land surface topography evolved. In lowland areas, rivers occupy wide (<1.3 km) and generally linear valleys with cliff-bounded valley walls and generally shallow valley bottoms. Here, rivers are commonly underfit and sit on a variety of substrates, from bedrock to boulders to sand, which has influenced river morphology. This varies from single to anabranching channels, straight to meandering reaches with isolated terrace fragments in places, boulder bars, and pool and riffle systems.

Slope sediment systems are relatively common within the Cederberg. These consist of debris flows or fans located at the base of steep slopes within the larger river valleys and are located at the front of deeply incised bedrock or boulder channels that feed from the adjacent uplands. Sediments within the debris fans are blocky and poorly sorted, and based on weathering properties were deposited episodically, likely by flash floods (Boelhouwers et al. 1999). In areas of sandstone weathering, loose surface blocks also litter bedrock slopes, or have accumulated as rockfall fans (Fig. 10.2c).

10.6 Sandstone Weathering

One of the most characteristic features of the Cederberg is the spectacularly sculpted sandstone formations (Fig. 10.4) which are products of chemical and physical weathering of the TMG sandstones. The distinctive red-orange staining along joints and on exposed surfaces is a result of the oxidation of iron compounds embedded within the rocks. Large-scale physical weathering features, resulting from changes in fracture intensity, rock strength as well as cementation and porosity–permeability patterns, are manifest in many iconic Cederberg landmarks such as the Maltese Cross (a residual isolated sandstone outcrop about 20 m in height, situated on the plateau below Sneeuberg) and the Wolfberg Arch (a remnant sandstone arch, 15 m in diameter, situated on the Wolfberg ridge line) (Fig. 10.4).

On a smaller scale, a striking feature of Cederberg rock surfaces is their apparent karst-like appearance, which differs from the much smoother Karoo sandstone surfaces discussed in Chap. 2. To our knowledge, there appears to be almost no reference to karstic morphologies in the Cederberg other than that by Marker (1996) who termed these features as ‘pseudokarst’. Despite the highly resistant nature of TMG sandstones, features such as cave complexes (e.g. the Stadsaal Caves) and the karren-like deformations evident on upland rock surfaces seem to suggest that these features resulted from chemical (i.e. solution) weathering processes.

Cederberg soils, the final weathering products, are often yellow-brown, unlike the more greyish soils typically associated with the CBF; this is because Cederberg soils are derived, in part, from the Pakhuis and Cedarberg Formations. Most of the soils of the Cederberg are highly leached acid sands, coarse-grained, nutrient-poor and low in moisture-retaining capacity, with the exception of small pockets of loamy soils derived from the Cedarberg Formation shales. Soils on rocky slopes are skeletal and very porous, thus contributing to the stark rocky landscape on which it is difficult for vegetation to become established.

10.7 Occupation of Cederberg Landscapes

The weathering/erosion products such as caves and overhangs contain numerous examples of San rock art (Fig. 10.4d), earmarking the Cederberg as an important geoheritage region. There is a wide range of evidence showing human occupation in the Cederberg from the Middle Stone Age onwards (Orton and Mackay 2008), including hunter-gathering activities, cattle herding and trading, and rock art (van Rooyen 2012). Furthermore, the rock shelters within which rock art is well displayed often preserve rock hyrax (*Procavia capensis*) middens which are useful archives of Quaternary palaeoenvironmental information (e.g. Chase et al. 2011; Quick et al. 2011). Recent stable carbon and nitrogen isotope records from such hyrax middens in the

Fig. 10.4 Well-known landmarks in the Cederberg: **a** Clanwilliam Cedars (*Widdringtonia cedarbergensis*) at De Rif, central Cederberg (*Photograph* L. Quick), **b** the Wolfberg Arch (*Photograph source* http://www.geckocreek.com/activities/activ_wolfberg_arch.jpg), **c** the Maltese Cross (*Photograph source* http://cederbergwine.com/blog/wp-content/uploads/Cape-Nature-Maltese-cross-11.jpg) and **d** San rock art near the Stadsaal Caves (*Photograph* L. Quick)

Cederberg have revealed the first unequivocal terrestrial manifestation of the Younger Dryas cool climate period in southern Africa (Chase et al. 2011).

10.8 Conclusions

The scientific (geologic, geomorphologic, ecological, biological and archaeological) significance and scenic value of the Cederberg mark the region as a prime geotourism and geoheritage centre within South Africa. Iconic Cederberg landforms such the Maltese Cross, the Wolfberg Cracks and the Wolfberg Arch can be accessed on foot via demarcated hiking trails (Lourens 2013a, b). The long-term protection of the area remains an important goal for conservation. Provisional government and Cederberg landowners entered into a voluntary agreement to conserve 182,000 ha of private and state land, forming the Cederberg Conservancy which has a strong focus on geotourism and the sustainable management of the environment. Access to the Cederberg's key geoheritage sites is managed by members of the Cederberg Conservancy and CapeNature. Map guides (Lourens 2013a, b) that are issued to hikers and visitors to the area showcase the key geological and geomorphological features of the Cederberg. Unlike the rest of southern Africa, the Cederberg is comprised of relatively young and folded litholgies which has generated a unique and contrasting landscape. Large-scale earth processes of folding, faulting and weathering/erosion have shaped the sandstones and shales of the Cederberg into a majestically rugged mountainous terrain.

References

Aldridge RJ, Gabbott SE, Theron JN (2001) The Soom Shale. In: Briggs DEG, Crowther PR (eds) Palaeobiology II. Blackwell Science, Hoboken, pp 340–342

Boelhouwers JC, Jager DF, de Joode A (1999) Application of relative-age dating methods to openwork debris flow deposits in the Cederberg Mountains, Western Cape, South Africa. S Afr Geogr J 81:135–142

Broquet CAM (1992) The sedimentary record of the Cape Supergroup: a review. In: de Wit MJ, Ransome IGD (eds) Inversion tectonics of the cape fold belt, Karoo and Cretaceous Basins of Southern Africa. A.A. Balkema, Rotterdam, pp 159–183

Campbell BM, Werger MJA (1988) Plant form in the mountains of the Cape, South Africa. J Ecol 76:637–653

CapeNature (2013) Cederberg Wilderness map and brochure. http://www.capenature.co.za/wp-content/uploads/2013/11/Cederberg-map-and-brochure.pdf. Accessed on 7 May 2013

Chase BM, Quick LJ, Meadows ME, Scott L, Thomas DSG, Reimer PJ (2011) Late-glacial interhemispheric climate dynamics revealed in South African hyrax middens. Geology 39:19–22

Cowling RM, Lombard AT (2002) Heterogeneity, speciation/extinction history and climate: explaining regional plant diversity patterns in the Cape Floristic Region. Divers Distrib 8:163–179

de Beer CH (1992) Structural evolution of the Cape Fold Belt syntaxis and its influence on syntectonic sedimentation in the SW Karoo Basin. In: de Wit MJ, Ransome IGD (eds) Inversion Tectonics of the Cape Fold Belt, Karoo and Cretaceous Basins of Southern Africa. A.A. Balkema, Rotterdam, pp 197–206

de Beer CH (1995) Fold interference from simultaneous shortening in different directions: the Cape Fold Belt syntaxis. J Afr Earth Sci 21:157–169

de Wit MJ, Ransome IGD (1992) Regional inversion tectonics along the southern margin of Gondwana. In: de Wit MJ, Ransome IGD (eds) Inversion Tectonics of the Cape Fold Belt, Karoo and Cretaceous Basins of Southern Africa. A.A. Balkema, Rotterdam, pp 15–21

du Toit AL (1926) The Geology of South Africa. Oliver and Boyd, Edinburgh 463 pp

du Toit AL (1954) The geology of South Africa, 3rd edn. Oliver and Boyd, Edinburgh 611 pp

Gresse PG, Theron JN (1992) The geology of the Worcester area: explanations of sheet 3319 (1:250 000). Geological Survey of South Africa, Pretoria

Hijmans R, Cameron SE, Parra JL, Jones PG, Jarvis A (2005) Very high resolution interpolated climate surfaces for global land areas. Int J Climatol 25:1965–1978

King LC (1963) South African scenery. A textbook of geomorphology, 3rd edn. Oliver and Boyd, Edinburgh, 308 pp

Lourens T (2013a) Map Guide Sneeuberg in the Cederberg, Western Cape of South Africa. Blue Mountain Design and Publishing, Cape Town

Lourens T (2013b) Map Guide Wolfberg in the Cederberg, Western Cape of South Africa. Blue Mountain Design and Publishing, Cape Town

Marker ME (1996) The karstic landscape. In: Lewis CA (ed) The geomorphology of the Eastern Cape South Africa. Grocott and Sherry Publishers, Grahamstown, pp 87–102

Mustart P, Juritz J, Makua C, van der Merwe SW, Wessels N (1995) Restoration of the Clanwilliam Cedar *Widdringtonia cedarbergensis*—the importance of monitoring seedlings planted in the Cederberg, South Africa. Biol Conserv 72:73–76

Orton J, Mackay A (2008) New excavations at Klein Kliphuis rock shelter, Cederberg Mountains, Western Cape, South Africa: the late Holocene deposits. S Afr Archaeol Bull 63:69–76

Quick LJ, Chase BM, Meadows ME, Scott L, Reimer PJ (2011) A 19.5 kyr vegetation history from the central Cederberg Mountains, South Africa: Palynological evidence from rock hyrax middens. Palaeogeogr Palaeoclimatol Palaeoecol 309:253–270

Rowe CD, Backerberg NR (2011) Discussion on: reconstruction of the Ordovician Pakhuis ice sheet, South Africa by HJ Blignault and JN Theron. S Afr J Geol 114:95–102

Rust IC (1967) On the sedimentation of the Table Mountain Group in the western Cape Province. Unpublished PhD Thesis, Stellenbosch University

Schulze RE (1997) Atlas of agro-hydrology and agro-climatology, WRC Report TT82/96. Water Resource Commission, Pretoria

Shone RW, Booth PWK (2005) The Cape Basin, South Africa: A review. J Afr Earth Sci 43:196–210

Sutcliffe OE, Dowdeswell JA, Whittington RJ, Theron JN, Craig J (2000) Calibrating the late Ordovician glaciation and mass extinction by the eccentricity cycles of Earth's orbit. Geology 28:967–970

Tankard A, Welsink H, Aukes P, Newton R, Stettler E (2009) Tectonic evolution of the Cape and Karoo basins of South Africa. Mar Pet Geol 26:1379–1412

Valsecchi V, Chase BM, Slingsby JA, Carr AS, Quick LJ, Meadows ME, Cheddadi R, Reimer PJ (2013) A high resolution 15,600-year pollen and microcharcoal record from the Cederberg Mountains, South Africa. Palaeogeogr Palaeoclimatol Palaeoecol 387:6–16

van Rooyen P (2012) Two southern African rock art sites as indicators of ancient migratory routes. Rock Art Res 29:179–186

Visser JNJ (1974) The table mountain group: a study in the deposition of quartz arenites on a stable shelf. Trans Geol Soc S Afr 77: 229–237

Young, GM, Minter, W, Theron, J (2004) Geochemistry and palaeogeography of upper Ordovician glaciogenic sedimentary rocks in the Table Mountain Group, South Africa. Palaeogeogr Palaeocl Palaeoecol 214:323–345

[illegible] *Zeitschrift für Geomorphologie* [illegible] Suppl. [illegible]

[illegible] geomorphology and plant diversity patterns [illegible] Journal of Biogeography [illegible]

[illegible]

Table Mountain: Wonder of Nature at the Foot of Africa

11

Michael E. Meadows and John S. Compton

Abstract

The geology in the vicinity of the city of Cape Town is exceptional for its spectacular scenery, distinctive geomorphology, the diversity of rock types and contacts on display, and its relation to the highly diverse natural fynbos vegetation. The iconic Table Mountain rises over 1,000 m above the city centre and forms the focus of the recently established Table Mountain National Park. The most obvious geomorphological features are the dramatic cliff faces and steep slopes that expose the three main rock types of the area: metamorphic rocks of the Malmesbury Group, igneous rocks of the Cape Granite Suite (including intrusions of dolerite dykes) and sedimentary rocks of the Table Mountain Group. Tectonism along with differences in weathering processes of different rock types has combined to give the distinctive flat-top of Table Mountain, the rounded, smooth slopes of the Tygerberg, Signal Hill and Paarl Rock, and the expansive Cape Flats in between.

Keywords

Cape Town • Seven Wonders of Nature • Cape Fold Belt mountains • Sandstone • Shale • Granite

11.1 Geographical and Environmental Setting

There are few more spectacular or iconic landscapes in South Africa than the conspicuous 1,000-m-high flat-topped mesa of Table Mountain framed by the nearby Devil's Peak and Lion's Head/Signal Hill (Figs. 11.1 and 11.2). The prominent massif rises steeply from the Atlantic Ocean and forms a distinctive landmark at the northern end of the Cape Peninsula. Its cliffs provide a backdrop to the city of Cape Town which, when juxtaposed with sea to the west and the low-lying Cape Flats to the east, makes for one of the most awesome urban locations in the world. This chapter explores the origins and context of Table Mountain.

Table Mountain is a small part of a much larger mountain chain known as the Cape Fold Belt that extends as far as Port Elizabeth (800 km to the east) and the Cederberg (300 km to the north; see Chap. 10). The cliff face of the mountain lies at the northern end of the Cape Peninsula, an outlying extension to the Cape Fold Belt. The most common rocks of the Cape Fold Belt are sandstones, and these form the rugged summits and exposed cliff faces of Table Mountain, although the underlying geology is more varied and the evolution of the landscape is complex.

The underlying geology of the south-western Cape (Theron et al. 1992) is expressed distinctively in the landscape. Table Mountain and adjacent promontories provide a good example of the relationship between bedrock composition, structure and topographic form. The late Precambrian Malmesbury Group shales are the oldest basement rocks of the region, and the constituent metamorphosed fine mudstones and sandstones are relatively soft and underlie much

M.E. Meadows (✉) · J.S. Compton
University of Cape Town, Cape Town
South Africa
e-mail: michael.meadows@uct.ac.za

J.S. Compton
e-mail: john.compton@uct.ac.za

S. Grab and J. Knight (eds.), *Landscapes and Landforms of South Africa*,
World Geomorphological Landscapes, DOI 10.1007/978-3-319-03560-4_11,

Fig. 11.1 The landscape of Table Mountain looking to the south-east and illustrating major geomorphological elements and features mentioned in the text. The *dashed line* is the original extent of Table Bay; the solid *white line* is the contact between the Cape Granite to the south and the Malmesbury shale to the north; the diagonally hatched line is the Graafwater Formation in depositional contact with either the granite or Malmesbury shale (*Photograph* M. Skinner; *The Rocks and Mountains of Cape Town*)

of the Cape Flats. Occasionally, however, the shales are more resistant due to intrusion and metamorphism by the Cape Granites and occur as steeply sloping but curvaceous hills, such as Mowbray Ridge and Signal Hill (Fig. 11.1). These hills are visible as a series of bulbous exposures on the northern side of Table Mountain and also as the Tygerberg (450 m high). It is from these hard, metamorphosed shales that aggregate is mined to pave the region's roads. The Cape Granites also weather to form rounded hills, such as the lower slopes of Lion's Head (Fig. 11.1) and more famously at Paarl Rock (40 km north-east of Cape Town).

Although the slopes of granite intrusions and metamorphosed Malmesbury sediments can be steep, it is only the highly resistant quartzose sandstones that produce steep cliff faces such as those above the Cape Town city bowl. These cliff faces are formed of hard, quartzitic sandstones that outcrop throughout the Cape Fold Belt and give rise to its rugged, high-relief topography through different geological and geomorphological processes.

11.2 Geological Evolution of Table Mountain

In order to more fully understand the Table Mountain landscape, it is necessary to consider the long-term geological evolution of the region, in particular because the mountain landforms are strongly determined by their underlying rocks. Lithification of Malmesbury marine sediments into shale rock was followed by intermediate intrusion of molten Cape Granite bodies which locally metamorphosed the host shales before cooling into the coarse crystalline granite that is now exposed at the surface. The contact between the two lithologies is exceptionally well exposed on the wave-cut platform at the Sea Point contact located in Bantry Bay. These deeply buried rocks rose to the surface to form a platform on which the mudstone and sandstones of the Table Mountain Group were deposited. These sediments, over 7 km thick (Compton 2004), were deposited between ~500 and 340 Ma in a subsiding basin (Tankard et al. 2009).

The Table Mountain Group, which makes up the lowermost part of the Cape Supergroup, is comprised of several sedimentary formations. The oldest of these (the Graafwater Formation) was deposited under an oxygen-rich atmosphere such that the iron has oxidised into deep maroon or rust coloured 'red beds' (Compton 2004) (Fig. 11.3). These beds have well preserved structures including cross-bedding, ripple marks, stream lineations and desiccation cracks (Theron et al. 1992). The contact between the Graafwater and the overlying quartz arenite sandstones of the Peninsula Formation is marked by a distinct change in slope where the more resistant quartzitic sandstones rise up as the impressive cliff faces of Table Mountain (Fig. 11.1). These sediments are ~550 m thick on Table Mountain but elsewhere are up to 1,800 m thick and consist almost entirely of well-sorted quartz sand (98 % by weight) with only small quantities of iron and manganese oxides and other minerals (Theron et al. 1992). The sandstones are uniformly light grey, medium to coarse-grained and thick bedded. Two depositional environments have been suggested for sandstone formation: a subtidal shelf during marine transgressions (Tankard and Hobday 1977) and prograding braided streams (Turner et al. 2011). Given the widespread occurrence of the sandstones, both environments may have occurred (Tankard et al. 2009, 2012). The overlying, relatively thin (40 m on Table Mountain) and

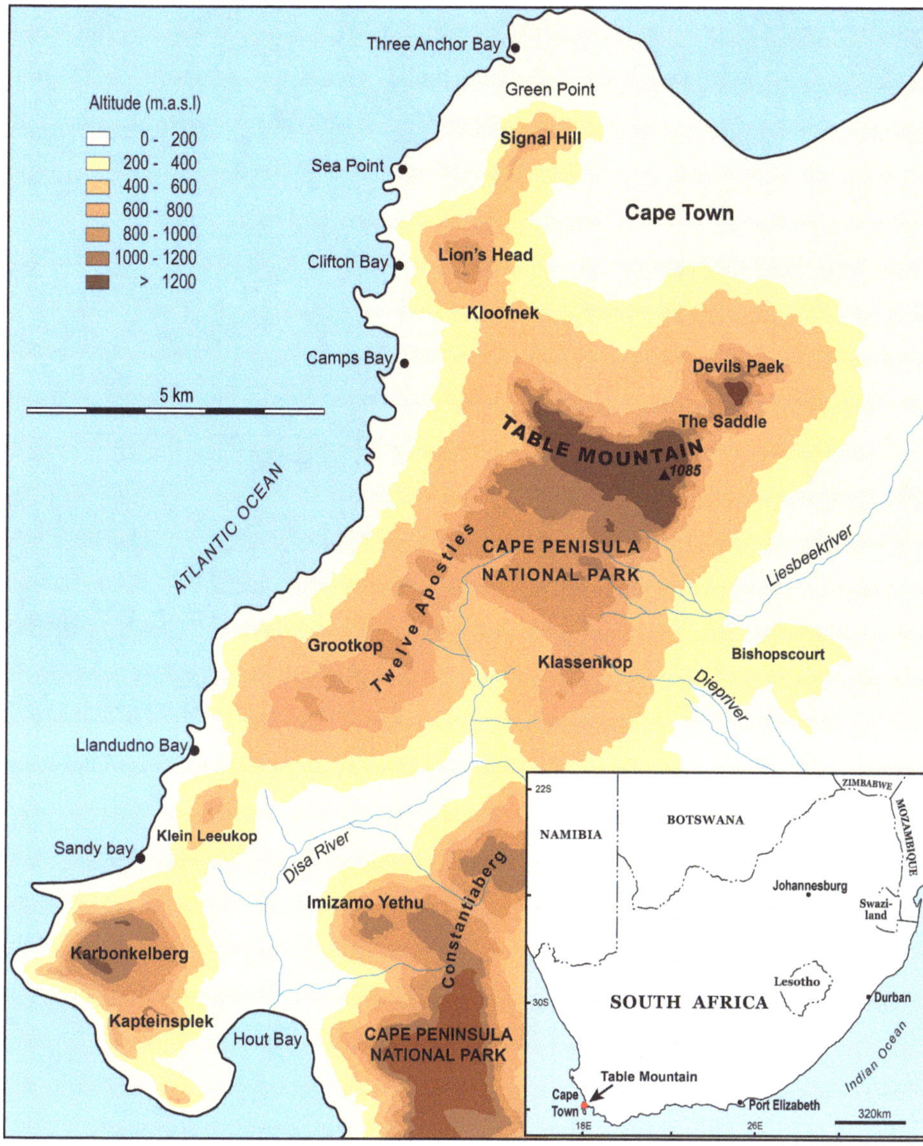

Fig. 11.2 Geographic and topographic setting of Table Mountain

discontinuous Pakhuis Formation contains faceted and striated pebbles within a diamictite (tillite) and is interpreted as a glacial deposit (Young et al. 2004; Turner et al. 2011). Massively bedded Pakhuis tillite is exposed at Maclear's Beacon, the highest point on Table Mountain (1,086 m).

11.3 Geomorphology of Table Mountain

The 1,000 m or so relief of Table Mountain is impressive and yet these rocks comprise merely the oldest, lowermost formations of the Cape Supergroup. Erosion of overlying sediments, which are preserved elsewhere in the Western Cape, coupled with large-scale tectonic activity has given rise to the characteristic flat-top of Table Mountain. A possible scenario for this geomorphic evolution is illustrated in Fig. 11.4. Overlying sediments, including the softer Cedarberg Formation shales, were selectively removed so that Table Mountain represents the remains of the synclinal trough of the ancient fold that extended across what is today the Cape Flats. Thus, the modern landscape is an inverted version of an earlier ancient landscape (Compton 2004).

Following initial tectonic folding during the formation of Gondwana and Pangaea and then the break-up of Pangaea, erosion gradually removed the overlying rocks to leave Table Mountain and its peninsula mountain chain, as well as the Piketberg and Riebeek Kasteel farther north, as isolated erosional remnants of a previously larger Cape Fold Belt. Given the extensive and intense tectonic activity that accompanied continent formation and break-up, it is remarkable that the Graafwater and Peninsula Formations have survived as near-horizontal layers within Table Mountain.

Fig. 11.3 Sedimentary rocks of the Graafwater Formation; note the *reddish brown* colouration that indicates oxidised iron-rich fine sandstones and mudstones (*Photograph* M. Meadows)

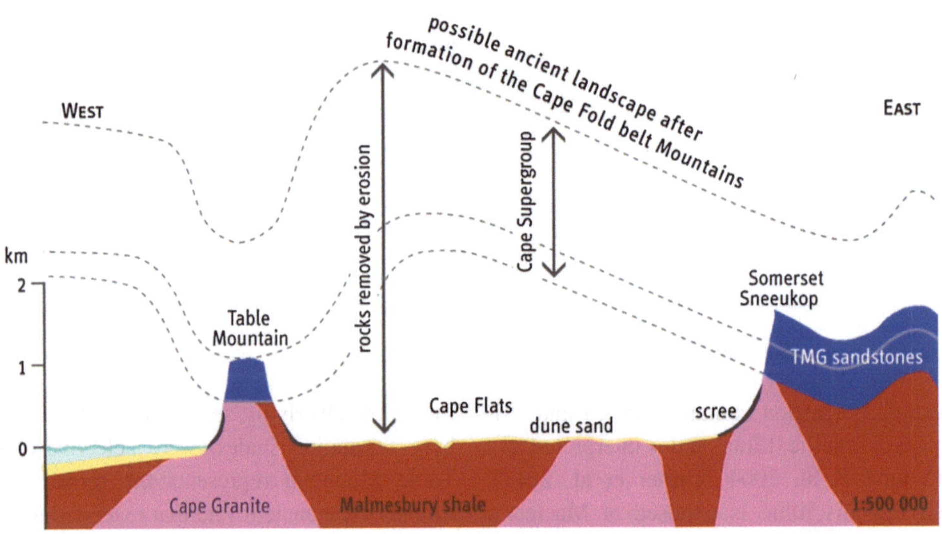

Fig. 11.4 Possible evolution of the landscape of Table Mountain and associated parts of the Cape Fold Belt. The almost horizontal sandstones of Table Mountain represent the syncline and the steeply dipping sandstones of the mountains to the east, the limb of a large fold that has since been eroded to expose the underlying shales. The modern geomorphic expression is, therefore, effectively an inverted version of the ancient landscape (Fig. 16 from *The Rocks and Mountains of Cape Town*)

Despite the absence of substantial folds, weathering of Table Mountain sandstones over millions of years has etched major gorges in the cliff faces. Several large clefts have also formed along faults within the sandstones, such as at the so-called Twelve Apostles above Camps Bay (Fig. 11.5). These gorges are associated with faults activated during Gondwana break-up and are long since inactive, but the small fault offset was sufficient to initiate erosion by stream run-off. During winter frontal storms, streams cascade off the mountain top and form temporary waterfalls that are capable of carrying substantial volumes of material including large boulders that are visible in the streambeds or within debris

Fig. 11.5 The 'Twelve Apostles' looking south from the Atlantic shoreline. Streams have cut steep-sided gorges into the hard Peninsula Formation sandstones; the underlying granite is exposed in the foreground (*Photograph* from *The Rocks and Mountains of Cape Town, Fig 17*)

Fig. 11.6 One of the many boulder deposits that cascade off the mountain (*Photograph* M. Meadows)

flow deposits (Fig. 11.6). Over many years, these streams have carved V-shaped gorges, which are host to the distinctive Afrotemperate forests, one of the few areas on the mountain with indigenous trees.

Debris flows and rockfalls (Fig. 11.7a, b) are also common below the cliff faces and testify to the intense geomorphic processes and associated geohazards to the urban infrastructure below. Rockfalls from Table Mountain cliffs

(a)

(b)

Fig. 11.7 a Evidence of a major rockfall in the sandstone cliffs on the east-facing slopes of the mountain. The more or less smooth 'scar' (which is approximately 100 m across and 50 m high) of the rockfall indicates that the event took place in the relatively recent geological past and is testimony to the fact that such processes are still active on Table Mountain. **b** The substantial talus slope below comprises the fragments of the large block that broke away, some of the individual boulders are the size of a double-decker bus (*Photographs* M. Meadows)

occur when the more easily eroded granite or Malmesbury shale is removed to produce oversteepened cliff faces. Episodic heavy rainfall or earthquakes cause the sudden release of large sandstone blocks and catastrophic rockfalls. The remains of past rockfalls are evident on the lower slopes as scattered, variably sized blocks of sandstones which may be preserved several 100 m distance from the modern cliff face. Rockfalls onto the road that connects the suburbs of Hout Bay and Noordhoek have necessitated major geoengineering interventions including a 'half tunnel' (Moolman 2005). Such features testify to ongoing mass movements on Table Mountain and the Cape Peninsula.

11.4 Table Mountain Landscapes: The Nature and Extent of Human Influence

The Cape Fold Belt exhibits distinctive fynbos vegetation that is associated with the nutrient-poor sandstones of this region. As with the geomorphology, the natural vegetation type is strongly influenced by underlying geology. Mountain fynbos communities are found on the shallow and nutrient-poor soils developed on the Table Mountain Group, although Afrotemperate forest patches occur in the steep-sided kloofs (mountain passes or gorges) on east-facing slopes. Malmesbury shales typically develop more subdued topography on which finer-grained and nutrient-rich soils have formed that favour renosterveld vegetation. The plant communities found on Table Mountain are fire-adapted and exhibit a remarkable degree of diversity and endemism, amplified by the spatial variability of microclimates induced by the diverse topography. With 2,285 plant species, the region is a 'hot spot' of global biodiversity (Myers et al. 2000). Many of the species present within the mosaic of mountain fynbos communities have only isolated, small populations, but these communities often exhibit a similar structure. Common species include shrubs belonging to the families Proteaceae and Ericaceae, geophytes include fires lilies (typically *Cyrtanthus ventricosus*) and, while grasses are relatively unusual, reeds of the Restionaceae are abundant across a wide range of habitat types.

The conservation status of many fynbos communities, and the Cape flora as a whole, has been markedly impacted by human activities including land use change for agriculture and urban development. Alien tree plantation and infestation and fire management problems also threaten the fynbos vegetation, especially under potential impacts of climate change. There have also been geomorphological consequences of human impact. Catchment hardening under urbanisation has dramatically altered the hydrological regime and created considerable flooding problems. Even though the catchment of Table Mountain as a whole is relatively small and given that there are several water supply reservoirs on the plateau top, flooding after heavy rain is a common problem in the low-lying Cape Flats. Plantations of pine and eucalypt species that once covered much of the lower slopes of the mountain following commercial afforestation during the late nineteenth and much of the twentieth centuries, especially on its northern and eastern flanks, are now being cleared as part of the 'Working for Water' government programme. The potential increase in surface run-off with associated risks of accelerated soil erosion requires careful monitoring. Soil erosion is not likely significant on steep sandstone cliffs or shallow (or absence of) sandstone soil, but it is significant on the Malmesbury shale and granite-derived soils on the lower slopes where there is major gullying.

11.5 Geoheritage

Table Mountain was declared a National Monument in 1958 (Rebelo et al. 2011) and, following long consultation, the Cape Peninsula National Park was declared in 1998 (Daitz and Myrdal 2009) and subsequently renamed 'Table Mountain National Park'. Although a major motivation behind the establishment of the park was its biodiversity, the origin of this biodiversity owes much to the diverse rock types and their associated soils. The mountain and its surroundings have long attracted visitors, but also scientists. Charles Darwin visited Cape Town in 1836, on the final leg of the five-year voyage of the *HMS Beagle,* and described some of the local physiographic features, including the 'hot' contact between the intruded granite and host Malmesbury shale exposed at Sea Point (Master 2012). Darwin's observations and arguments for the magmatic origin of the granite, a view that is accepted today, contributed to contemporary discussions on the origins of igneous rock, whether they formed from molten magma or precipitated cold out of sea water.

One of the main attributes of Table Mountain from a geoheritage perspective is its accessibility; not only are there spectacular views *of* the mountain from many parts of the city and beyond, there are remarkable vistas *from* the mountain too. The cable car, established in 1929 and operated by the Table Mountain Aerial Cableway Company Limited, provides effortless direct access to the summit. It is no surprise that in 2011, Table Mountain was voted one of the new seven natural wonders of the world in a global poll.

Given that the natural environment is arguably Cape Town's most important asset, the fact that it has the highest population growth rate of any city in South Africa emphasises the imperative of appropriate, comprehensive and integrated plans to ensure its protection (Holmes et al. 2012). From a biodiversity perspective, the situation has been described as 'perilous' (Rebelo et al. 2011), but conservation planning needs to go beyond the protection of the region's biological heritage to adopt a holistic approach and consider the Table Mountain landscape in its entirety, including its geological and geomorphological features.

References

Compton JS (2004) The rocks and mountains of Cape Town. Double Storey, Cape Town, 112 pp

Daitz D, Myrdal B (2009) Table Mountain National Park. In: Suich H, Child B, Spenceley A (eds) Evolution and innovation in wildlife conservation. Earthscan, London, pp 325–340

Holmes PM, Rebelo AG, Dorse C, Wood J (2012) Can Cape Town's unique biodiversity be saved? Balancing conservation imperatives and development needs. Ecol Soc 17. http://dx.doi.org/10.5751/ES-04552-170228

Master S (2012) Darwin as a geologist in Africa—dispelling the myths and unravelling a confused knot. S Afr J Sci 108:1–5

Moolman S (2005) Rehabilitation of Chapman's Peak Drive: a triumph for South African engineering. Civil Eng 13:5–7

Myers N, Mittermeier RA, Mittermeier CG, da Fonseca GAB, Kent J (2000) Biodiversity hotspots for conservation priorities. Nature 403:853–858

Rebelo AG, Holmes PM, Dorse C, Wood J (2011) Impacts of urbanization in a biodiversity hotspot: conservation challenges in Metropolitan Cape Town. S Afr J Bot 77:20–35

Tankard AJ, Hobday DK (1977) Tide-dominated back barrier sedimentation, early Ordovician Cape Basin, South Africa. Sed Geol 18:135–139

Tankard A, Welsink H, Aukes P, Newton R, Stettler E (2009) Tectonic evolution of the Cape and Karoo basins of South Africa. Mar Pet Geol 26:1379–1412

Tankard A, Welsink H, Aukes P, Newton R, Stettler E (2012) Geodynamic interpretation of the Cape and Karoo basins, South Africa. In: Roberts DG, Bally AW (eds) Phanerozoic passive margins, cratonic basins and global tectonic maps, Elsevier, pp. 869−945

Theron JN, Gresse PG, Siegfried HP, Rogers J (1992) The geology of the Cape Town area. Geological Survey of South Africa, Pretoria 140 pp

Turner BR, Armstrong HA, Holt P (2011) Visions of ice sheets in the early Ordovician greenhouse world: evidence from the Peninsula Formation, Cape Peninsula, South Africa. Sed Geol 236:226–238

Young G, Minter WEL, Theron JN (2004) Geochemistry and palaeogeography of upper Ordovician glaciogenic sedimentary rocks in the Table Mountain Group, South Africa. Palaeogeogr Palaeoclimatol Palaeoecol 214:323–345

The Cape Winelands

12

Michael E. Meadows

Abstract

The major wine-producing region of South Africa is located in the south-western part of the country and widely known as the Cape Winelands. The region is geologically characterised by rocks of the Malmesbury, Cape Granite and Table Mountain Groups. Vineyards are located on soils developed under all three of these major geological substrates where slope conditions allow, as well as on the sand and gravel plains of the rivers that drain the region. The most striking geomorphological feature is the contrast between relatively gently rolling plateau country of the Swartland underlain by Malmesbury shales and Cape Granite, and the rugged topography of the Table Mountain Group sandstone-dominated Cape Fold Belt mountains to the east. Combinations of climate, geology, slope and soil factors, coupled with the efforts of the winemakers, give rise to contrasting *terroirs* in the region which favours the production of different styles, character and quality of wine produced. The winelands represent a suite of cultural landscapes that justify significant conservation efforts both for their historical and natural resource value.

Keywords

Western Cape • Vineyards • Terroir • Geology • Soils

12.1 Geographical and Environmental Setting

From its modest beginnings in the Company Garden on the slopes of Table Mountain in the eighteenth century, the wine-growing areas of the Western Cape have expanded markedly and vineyards now occupy more than 100,000 ha across the country such that South Africa is now the eighth largest wine-producing nation, representing about 3.6 % of global production volume (Van Zyl 2013). The extent of the vineyards, which stretch as far east as Plettenberg Bay and as far north as Upington (Fig. 12.1), is such that it is meaningless to attempt a description of the diversity of landforms and landscapes across the entire area. However, the range of main landscape types typical of the wineland region as a whole occurs within the south-western part of the region, extending as far north as Darling and north-eastwards to Paarl. This was also the logic employed by Talbot (1971) in his comprehensive geographical essay on the south-western part of what is today the Western Cape.

The broader area encompasses landscapes, the character of which are in essence determined by the interplay of geology, climate and land use, and given identity in their regional names as designated by European settlers in the seventeenth and eighteenth centuries (Fig. 12.2). The sandy soils of the coastal plain offer conditions considerably less suitable for the growing of grapes, so the two major regions of wine growing are the Swartland and the Voorberg. As Talbot (1971) notes, eastwards from the Sandveld, there is a broad, undulating plain cut into the Proterozoic (Precambrian) Malmesbury Group shales and Cape Granites that has, especially within the last decade or so, evolved from an area typified by wheat crops to one that is now important for

M.E. Meadows (✉)
University of Cape Town, Cape Town, South Africa
e-mail: michael.meadows@uct.ac.za

S. Grab and J. Knight (eds.), *Landscapes and Landforms of South Africa*,
World Geomorphological Landscapes, DOI 10.1007/978-3-319-03560-4_12,

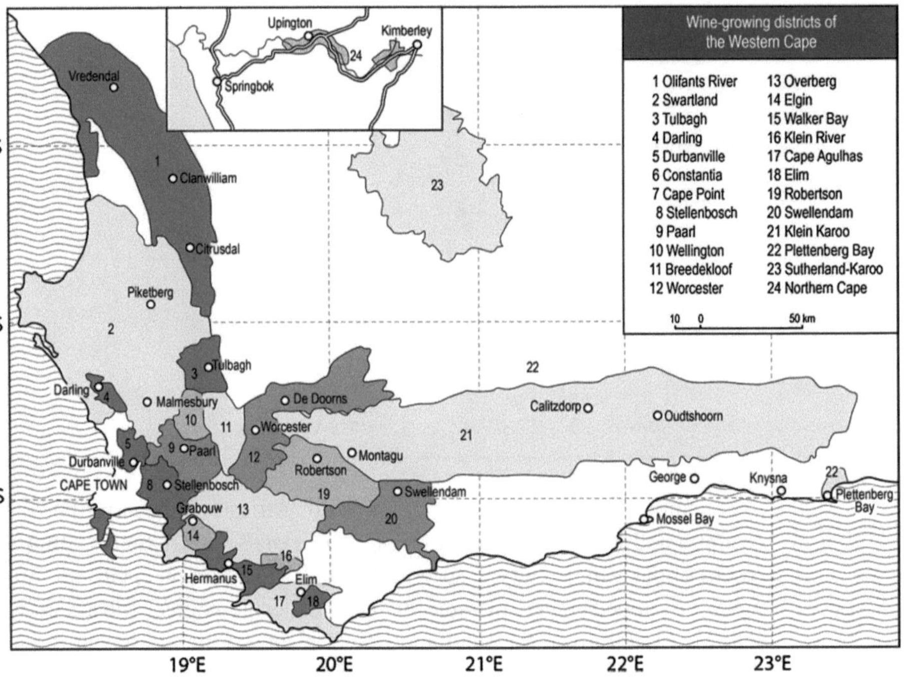

Fig. 12.1 The winelands of South Africa (map from Platter Guide 2013, p 532)

viticulture (Halpern and Meadows 2013). The eastern margins of this area are marked by outliers of the Cape Fold Belt mountains, and it is these that manifest as the most significant wine-producing parts of the region. Accordingly, there are really two major geological, geomorphological and soil combinations that characterise the winelands of the region, viz. (a) landscapes of subdued relief characterised by a relatively gentle undulating plateau with occasional granite intrusions that emerge as rounded hills, and (b) areas of high relief, rugged topography and shallow quartzitic soils developed on the sandstones of the fold belt mountains.

Key factors in wine quality include the character of the slopes and soils in which the vines are cultivated. This is very largely determined in the Cape Winelands by geological substrate, since the Malmesbury Group rocks have weathered into typically fine-grained and relatively nutrient-rich soils developed on slopes that rarely exceed 5 %. This stands in stark contrast to the situation on the weathering-resistant sandstones of the Table Mountain Group with its distinctively steep slopes and precipitous cliffs at the base of which are rocky screes, and even in the relatively gently sloping land of the major river valleys, such as the Breede and Berg, soils are very shallow, stony and always oligotrophic.

Of course climate conditions also play an important role in vine production. The south-western Cape in general is characterised climatically by the winter extension of transient rain-bearing westerlies (e.g. frontal depressions, ridging anticyclones and coastal low-pressure systems) accounting for the seasonality of precipitation (Chase and Meadows 2007). Due to the existence of the cold Benguela Current off the west coast, which exerts a strong stabilising influence on the air masses, there is marked summer aridity in the region. There are well-developed precipitation gradients associated with latitude, since areas to the north are considerably more xeric, and with altitude, since places at higher elevation are subject to wetter conditions. Mean annual rainfall at Stellenbosch in the south is 742 mm, at Malmesbury 460 mm and at Moorreesburg 386 mm (data courtesy of South African Weather Service). The marked spatial and temporal variation in rainfall in the region is an important factor in vineyard management and wine production.

12.2 Landscape

The relationship between underlying geology and topography is strikingly obvious in this region, and with the exception of areas of very subdued relief or where there is a covering veneer of geologically more recent aeolian or marine sand, the shape of the landforms is nearly always a very strong indication of the underlying parent material. The nature of the association is strongly evident in Fig. 12.3 (after Talbot 1971).

The Swartland, through a combination of geology, topography, soils and environmental and cultural history, has a distinctive landscape characterised by relatively

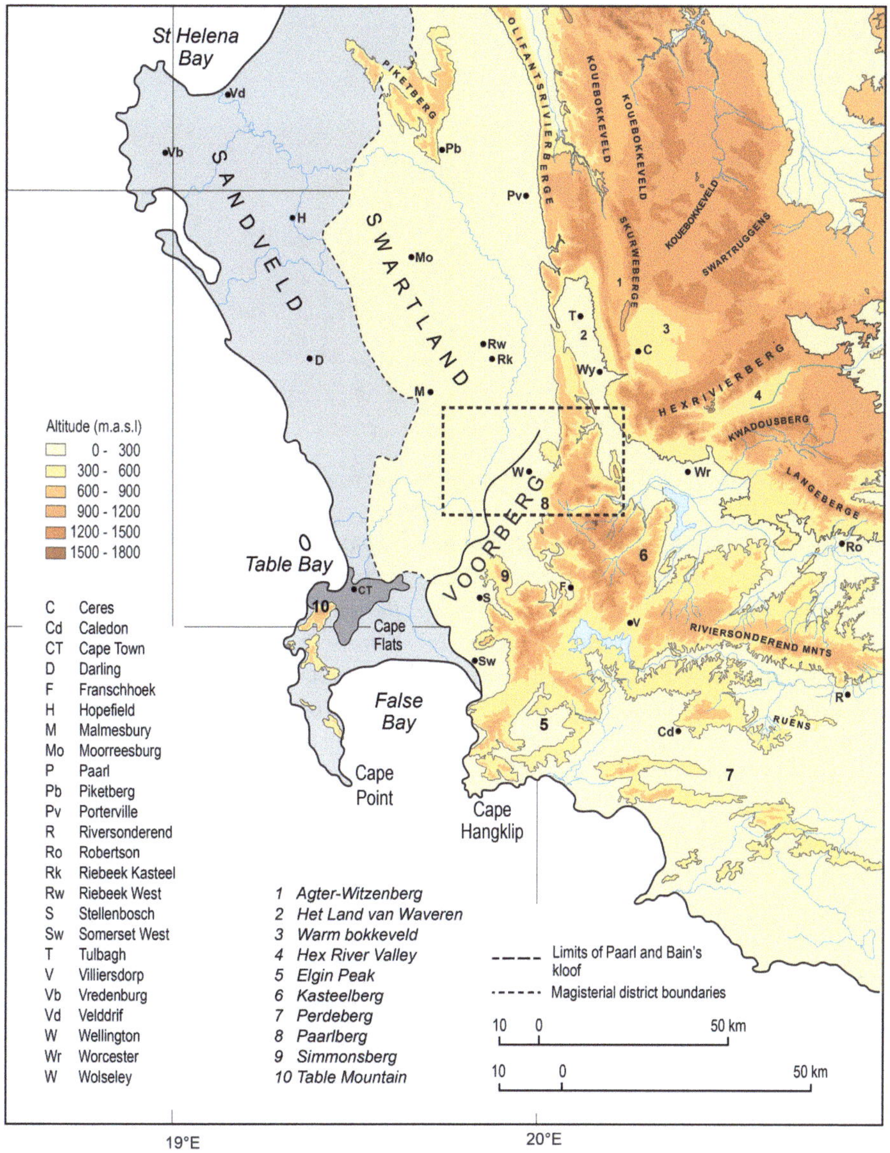

Fig. 12.2 The physiography of the Western Cape winelands (Talbot 1971; Fig. 12.1)

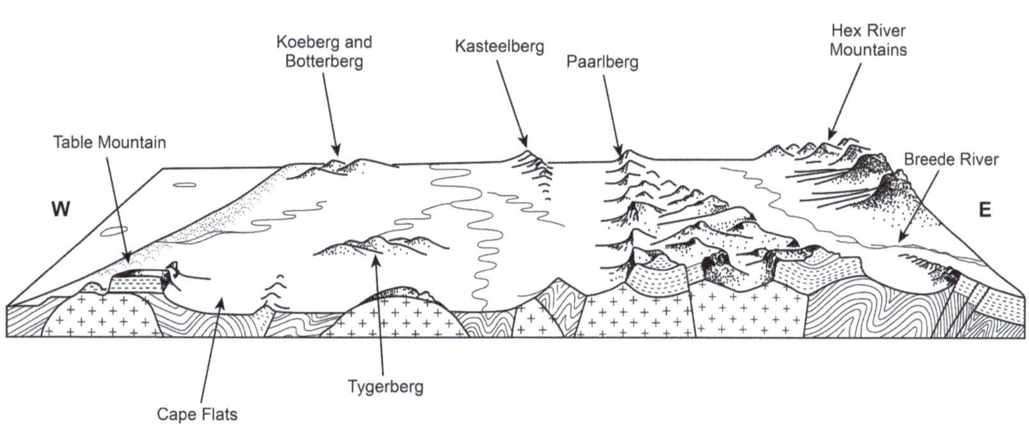

Fig. 12.3 The association of underlying geology and relief in the south-western Cape (Talbot 1971; Fig. 12.3)

Fig. 12.4 The undulating topography of the Swartland (*Photograph* M. Meadows)

subdued relief. Geologically, it is dominated by Precambrian metamorphosed shales and fine sandstones of marine origin known as the Malmesbury Group (Tankard et al. 1982). The generally undulating lowland (150–500 m) platform is punctuated by occasional broad granitic masses and steeper sided inselbergs of Table Mountain Group sandstones. The rivers that drain across the Swartland from the fold mountains to the east are incising themselves into the old marine floor that survives as the relatively level surface.

Where the intruded granites have been effective in hardening the otherwise softer shale, these have pronounced rounded, dome-shaped hills, such as the Tygerberg and Perdeberg (Fig. 12.4). Relatively shallow, brownish sandy loam duplex soils (Lambrechts 1998) are developed on Malmesbury rocks, and these are disposed to cracking after heavy rain and regarded as prone to sheetwash (Talbot 1947; Meadows 2003). In places, on footslopes in particular, the colluvium has accumulated to greater depths and the duplex nature of the soils has meant that they have been susceptible to accelerated erosion (Meadows 2003).

The most prominent geomorphic feature of what the early European settlers referred to as *die Gebergtes Afrikas* (Talbot 1971) is the belt of rugged quartzitic sandstone mountains that trends more or less north–south and rise in elevation to $\sim$2,000 m a.s.l. These mountains are comprised of a stack of Ordovician sedimentary sandstones and shales, viz. the Table Mountain Group, with a thickness exceeding 7,000 m (Theron et al. 1992; Compton 2004). They have been extensively folded and contorted by the tectonic forces that resulted in the initial formation and subsequent disaggregation of Gondwanaland (Shone and Booth 2005). The intensely folded mountains here are dissected by narrow gorges (known as *kloofs* in South Africa) so that the appearance is one of a complex of intersecting ranges. As Talbot (1971) notes, '*…the maze of mountains was named piecemeal, features and landmarks one by one receiving names, some descriptive and some recalling now vanished fauna, forgotten pioneers, and occasionally nostalgic memories—toponyms such as Sneeukop and Witteberge—the mountain system has no generally accepted inclusive name*' (p. 11). The resultant topography is distinctively craggy, and the lack of significant soil development on these hard, weathering-resistant sandstones has left the mountains looking spectacularly stark, barren and grey. Differential erosion in some of the units, for example the weaker shales above the tillite in the Table Mountain Group, has been selectively weathered, resulting in the undermining of the steep faces and producing a characteristic 'head and shoulders' geomorphology of some of the peaks. Rockfalls and scree or talus slopes are common features of the steeper cliff faces, while lower down, anastomosing rivers have produced a mantle of sands and gravels sculpted into low terraces that have proven particularly attractive as sites to plant vineyards.

12.3 Geomorphology and Cape Wines

The relationship between wine and geology, in terms of the productivity and quality of particular cultivars, is intuitively a strong one since the underlying rock influences soil type, permits or constrains root penetration and assists or hinders drainage (Bargmann 2003; Huggett 2006). Conradie et al. (2002) demonstrate how the combination of environmental factors, and not simply geology, influences varietal performance and character. Geomorphology too is an important factor in determining wine quality. Slope, in terms of altitude, aspect and angle, is a key element of the 'terroir' of a wine (van Leeuwen and Seguin 2006). The concept of terroir has been much vaunted and, indeed, debated but has clearly now become '*…a buzzword of the wine industry, mandatory in practically everything written…about wine including the label on the bottle, often without understanding*' (Haynes 1999, p. 191). The word is French, and there is probably no absolute equivalent in English, but it has a number of

Fig. 12.5 Vineyard terraces near Franschhoek (*Photograph* M. Meadows)

essentially geographical attributes of which geology and soil—and geomorphology for that matter—are very prominent (Townsend 2011). In fact, the relationship between wine grapes and geomorphology is such that it is common for wine farmers to directly manipulate topography on a mesoscale through ploughing and/or on a much larger scale through the construction of terraces, especially on steeper slopes (Fig. 12.5). Therefore, vineyard landscapes—and those of the south-western Cape are no exception—are in important ways 'anthropogeomorphic' products of human manipulation of geomorphology (Townsend 2011).

It is now accepted that the complex interplay between climate, geology, soil, geomorphology, grape cultivar and wine production methods plays a significant role in wine style and quality. There are numerous examples of how various components of the physical landscape may vary across the Cape Winelands and how these, in turn, influence wine character, style and quality (Carey et al. 2008). Slope aspect and position relative to the coast is a geomorphic characteristic that, through its influence on the nature and intensity of sea breezes, can have a marked influence on viticulture (Bonnardot et al. 2005). Detailed effects of soil and drainage characteristics have also been demonstrated, for example in relation to the production of Sauvignon Blanc and Cabernet Sauvignon varietals in the Helderberg area (Shange and Conradie 2012a, b). Topography also plays an important role, via aspect and slope position, in relation to temperature, for example hollows on south-facing slopes in particular may act as frost pockets and significantly influence vine performance in such localities. Not surprisingly, therefore, wine-growing areas within the south-western Cape that contrast in their physical environmental attributes give rise to contrasting terroirs and are characterised by, or better known for, different wine styles. It would be over-simplistic, given the geological and geomorphic diversity of the region as a whole, to summarise particular areas as restricted to or even typified by particular wine styles, but the geological and geomorphic influence on wine is nevertheless very significant. For example, parts of the Robertson region, with its shale-derived and alluvial soils, are renowned for white wines, whereas the Simonsberg, near Stellenbosch (the 'wine capital' of South Africa; Platter 2013), is more strongly associated with red wines produced in shallow and stony soils developed on the nutrient-poor Table Mountain Group sandstones. The geological age of the underlying rocks perhaps lends the Cape wines a special character as even immature Quaternary soils and landforms may determine the character of terroir and so influence the character of the resultant wine (Costantini et al. 2012).

12.4 The Nature and Extent of Human Influence

The Western Cape has a long history of human occupation, and archaeological studies at Pinnacle Point near Mossel Bay have led to the suggestion that parts of the region may have been the origin location for the lineage that leads to all modern humans, which appeared between 100,000 and 200,000 years ago in Africa (Brown et al. 2009). The first

Fig. 12.6 Typical winelands scenery in the Voorberg, the juxtaposition of dramatic cliff faces and vineyards is especially striking (*Photograph* M. Meadows)

behaviourally modern humans were using fire to work stone tools as early as 174,000 years ago, but their impacts on the environment remain uncertain. Much better known is the impact of colonial settlement (late seventeenth century onwards) on agriculture, and this appears to have been profound as local vegetation was cleared to make way for agriculture on a large scale with resultant geomorphic impacts including accelerated soil loss through wind erosion, especially in more exposed coastal localities of the Sandveld (Baxter and Meadows 1999). Quite substantial areas of the Western Cape appear to be susceptible to land degradation, perhaps even desertification.

Of these areas, the Swartland has been subject to significant levels of land degradation in the past, manifesting as widespread gully erosion in the late 1930s (Talbot 1947). During the 1940s, the region was described as on the verge of economic collapse due to the severity of soil erosion, but concerted soil conservation and education efforts under the political dispensation of the time appear to have averted that scenario (Meadows 2003). Notwithstanding this, the Swartland has been virtually completely transformed for agriculture and only tiny remnants of the once widespread natural vegetation (Renosterveld) remain (Newton and Knight 2005). The region now faces the combined challenges of potentially rapid climate change under a considerably altered socio-economic and political order and a dynamic land use situation which has seen substantial areas traditionally under wheat being converted to vineyards with, as yet, unknown environmental consequences (Halpern and Meadows 2013).

Mainly because the slopes are too steep for development and the soils too nutrient-poor and shallow for widespread agriculture, the human impact consequences on the uplands of the Voorberg underlain by Table Mountain Group sandstones have been somewhat less pronounced. Nevertheless, it is these footslopes—and the alluvial plains emerging from them—that provided the stage for the expansion of vineyards into the interior from their initial establishment on the slopes of Table Mountain. This has had negative effects on natural vegetation and associated animal communities rather than on geomorphology per se, but the fact that the region is at the core of one of the world's major biodiversity hotspots (Myers et al. 2000) adds considerable impetus to the necessity for adequate conservation measures to be in place. Fire management is a key element of this conservation planning. Even the rather shallow soils on steep slopes may be susceptible to instability following a hot burn when geomorphic processes such as debris flows and rockfalls are significantly more common.

12.5 Geoheritage

The viticultural landscapes of the south-western Cape are key resources both from the direct economic perspective as wine growing is, of itself, a major industry in the region, but also in the indirect sense as they are a major tourist attraction. As Costantini et al. (2012) have observed, in wine-growing areas of Europe, such as the Montepulciano territory of Italy, the vineyard soils are products of human activity and formed as a consequence of a unique combination of natural and anthropogenic processes so that they themselves can rightly be considered cultural heritages. Wine is arguably the key tourism attraction in the Western Cape; indeed, Bruwer and Alant (2009) refer to the almost intoxicating attraction of the product and that '... *the decision to engage in wine tourism is generally impulsive, even spurious, the visit duration short and the motivation guiding the visitors' behaviour predominantly hedonic in nature*' (p 235). However, the spectacular landscapes of the Cape vineyards, with their dramatic backdrop of fold mountains, are certainly an enduring image (Fig. 12.6), one that is much used in the promotional material produced by the wine estates and surely an additional major allure.

The winelands are in essence a cultural landscape and certainly warrant conservation. This has been recognised formally in some cases, for example in the Idas Valley near Stellenbosch which has been protected as a heritage site since 1976 (Pistorius and Todeschini 2004). In many ways, this area typifies both the glory and the tragedy of South African history since the valley was home firstly to *San* hunter-gatherer and then *Khoi* herder populations before the European colonists forcibly displaced them or enslaved them from 1682 onwards. There is evidence of human landscape modifications over a long period in response to historical factors that have influenced agriculture in the region and that are recognised for their heritage value. While the geomorphology or geology are of themselves not subject to particular conservation efforts, the '...*magnificent natural setting, comprising dramatic mountain wilderness, rolling hills and gently sloping valley lands, streams and springs, gravelly and rich alluvial soils and associated diverse flora (fynbos) and fauna*' (Pistorius and Todeschini 2004, p. 68) are significant natural assets that provide the physical environmental motivation to conserve. There have also been rigorous efforts to conserve the biodiversity of the fynbos and associated animal communities through the establishment of the *Biodiversity and Wine Initiative* (BWI). The basic principle of the BWI, which is a partnership between the vineyards and various conservation organisations, including the Worldwide Fund for Nature (WWF), is to at least match each hectare under vineyard with a hectare of natural vegetation committed to conservation.

References

Bargmann CJ (2003) Geology and wine 7: geology and wine production in the coastal region, Western Cape Province, South Africa. Geosci Can 30:161–182

Baxter A, Meadows ME (1999) Evidence for Holocene sea level change at Verlorenvlei, Western Cape, South Africa. Quatern Int 56:65–79

Bonnardot V, Planchon O, Cautenet S (2005) Sea breeze development under an offshore synoptic wind in the south-western Cape and implications for the Stellenbosch wine-producing area. Theoret Appl Climatol 81:203–218

Brown KS, Marean CW, Herries AIR, Jacobs X, Tribolo C, Braun D, Roberts DL, Meyer MC, Bernatchez J (2009) Fire as an engineering tool of early modern humans. Science 325:859–862

Bruwer J, Alant K (2009) The hedonic nature of wine tourism consumption: an experiential view. Int J Wine Bus Res 21:235–257

Carey VA, Saayman D, Archer E, Barbeau G, Wallace M (2008) Viticultural terroirs in Stellenbosch, South Africa. I: the identification of natural terroir units. Journal Internationale des Science de la Vigne et du Vin 42:169–183

Compton JS (2004) The rocks and mountains of Cape Town. Double Storey, Cape Town, 112 pp

Conradie WJ, Carey VA, Bonnardot V, Saayman D, van Schoor LH (2002) Effect of different environmental factors on performance of sauvignon blanc grapevines in the Stellenbosch/Durbanville districts of South Africa. 1: geology, soil, climate, phenology and grape composition. S Afr J Enol Vitic 23:78–91

Costantini EAC, Bucelli P, Priori S (2012) Quaternary landscape history determines soil functional characters of terroir. Quatern Int 265:63–73

Chase BM, Meadows ME (2007) Late Quaternary dynamics of southern Africa's winter rainfall zone. Earth Sci Rev 84:103–138

Halpern AWG, Meadows ME (2013) Fifty years of land use change in the Swartland, Western Cape, South Africa: characteristics, causes and consequences. S Afr Geogr J 95:38–49

Haynes SJ (1999) Geology and wine 1: concept of terroir and the role of geology. Geosci Can 26:190–194

Huggett JM (2006) Geology and wine: a review. Proc Geol Assoc 117:239–247

Lambrechts JJN (1998) Soils of the Tygerberg region. In: Du Plessis NM (ed) The Tygerberg. Tafelberg Press, Cape Town, pp 91–111

Meadows ME (2003) Soil erosion in the Swartland, Western Cape Province, South Africa: implications of past and present policy and practice. Environ Sci Policy 6:17–28

Myers N, Mittermeier RA, Mittermeier CG, da Fonseca GAB, Kent J (2000) Biodiversity hotspots for conservation priorities. Nature 403:853–858

Newton I, Knight IP (2005) The use of a 60-year series of aerial photographs to assess local agricultural transformations of West Coast Renosterveld, an endangered South African vegetation type. S Afr Geogr J 87:18–27

Pistorius P, Todeschini F (2004) Idas valley as an example of the cultural landscape of the Cape Winelands. http://www.icomos.org/studies/viticoles/viticole8.pdf. Accessed 21 Nov 2013

Shange LP, Conradie WJ (2012a) Effects of soil parent material and climate on the performance of *Vitis vinifera* L. cvs. sauvignon blanc and cabernet sauvignon—part I: soil analysis, soil water status, root system characteristics, leaf water potential, cane mass and yield. S Afr J Enol Vitic 23:161–173

Shange LP, Conradie WJ (2012b) Effects of soil parent material and climate on the performance of *Vitis vinifera* L. cvs. sauvignon blanc and cabernet sauvignon—part II: climate, leaf analysis, juice analysis and wine quality. S Afr J Enol Vitic 23:174–183

Shone RW, Booth PWK (2005) The Cape Basin, South Africa: a review. J Afr Earth Sc 43:196–210

Talbot WJ (1947) Swartland and Sandveld. Oxford University Press, Cape Town

Talbot WJ (1971) South Western Cape Province: Paarl and Bain's Kloof Sheets. South African Geographical Society, 50 pp

Tankard AJ, Jackson MPA, Eriksson KA, Hobday DK, Hunter DR, Minter WEL (1982) Crustal evolution of Southern Africa: 3.8 billion years of earth history. Springer, New York, 523 pp

Theron JN, Gresse PG, Siegfried HP, Rogers J (1992) The geology of the Cape Town Area. Geol Surv S Afr, Pretoria, 140 pp

Townsend CG (2011) Viticulture and the role of geomorphology: general principles and case studies. Geogr Compass 5:750–766

Van Zyl P (ed) (2013) Platter's South African wines 2013. John Platter SA Wine Guide (Pty) Ltd, Hermanus, 626 pp

Van Leeuwen C, Seguin G (2006) The concept of terroir in viticulture. J Wine Res 17:1–10

The Kruger National Park: Geology and Geomorphology of the Wilderness

13

Morris Viljoen

Abstract

The Kruger National Park is an outstanding wildlife site and ecological resource where ecosystems and their functions are strongly influenced by underlying geology, geomorphology, soils, climate and water resources. This chapter outlines the general geology and geomorphology of the Kruger area. Landscape evolution and landform development during the Cenozoic has also been strongly controlled by the varied geological formations present in Kruger. Landscape development has included the exposure of bedrock surfaces by uplift and erosion, the accumulation of weathered products to form soils and the subsequent transportation of alluvial sediments within the riparian zone of Kruger's rivers. The interrelationships between these factors and their role in ecosystem development and cultural landscape features are examined.

Keywords

Geology • Geomorphology • Rock weathering • Soils • Rivers

13.1 Introduction

The Kruger National Park (here termed Kruger) is probably the best known national park in southern Africa, is also internationally well marketed and is an important tourist location. Kruger is part of the Greater Limpopo Transfrontier Park, comprising $\sim$35,000 km^2 in South Africa and adjoining regions of Mozambique and Zimbabwe (Fig. 13.1). Kruger is located in the north-eastern portion of the 'Lowveld' (i.e. low ground) and belongs to the 'lowveld geomorphic province' of South Africa (Partridge et al. 2010). The region has a relatively low elevation with a gentle east-dipping land surface and varies between 300 and 840 m a.s.l., the highest summit being Khandizwe Mountain in the south-western sector of the park (Fig. 13.1).

The region is best known internationally for its biodiversity and abundant and varied wildlife including, among others, the 'big five' of lion, African elephant, rhino (white and black), leopard and Cape buffalo. Consequently, the region has a strong emphasis on sustainability and ecosystem management [e.g. Southern African Wildlife College (SAWC)]. The Kruger region is, however, also of interest with respect to its varied underlying geology, geomorphology and soils which, together with climate, have exerted significant controls on ecosystem properties, processes and functions (Gertenbach 1983; Venter 1986; Mutanga et al. 2004). In addition, there have been considerable anthropogenic influences on the fluvial geomorphology of Kruger, primarily through the construction of dams across rivers and streams. This chapter focuses on the wider relationships between geology, geomorphology and landscapes of Kruger, with reference to the region's long-term landscape evolution. A number of major landscape types, largely determined by the varied underlying geology, are recognised and discussed.

M. Viljoen (✉)
University of the Witwatersrand, Johannesburg, South Africa
e-mail: morris.viljoen@vmic.co.za

S. Grab and J. Knight (eds.), *Landscapes and Landforms of South Africa*,
World Geomorphological Landscapes, DOI 10.1007/978-3-319-03560-4_13,

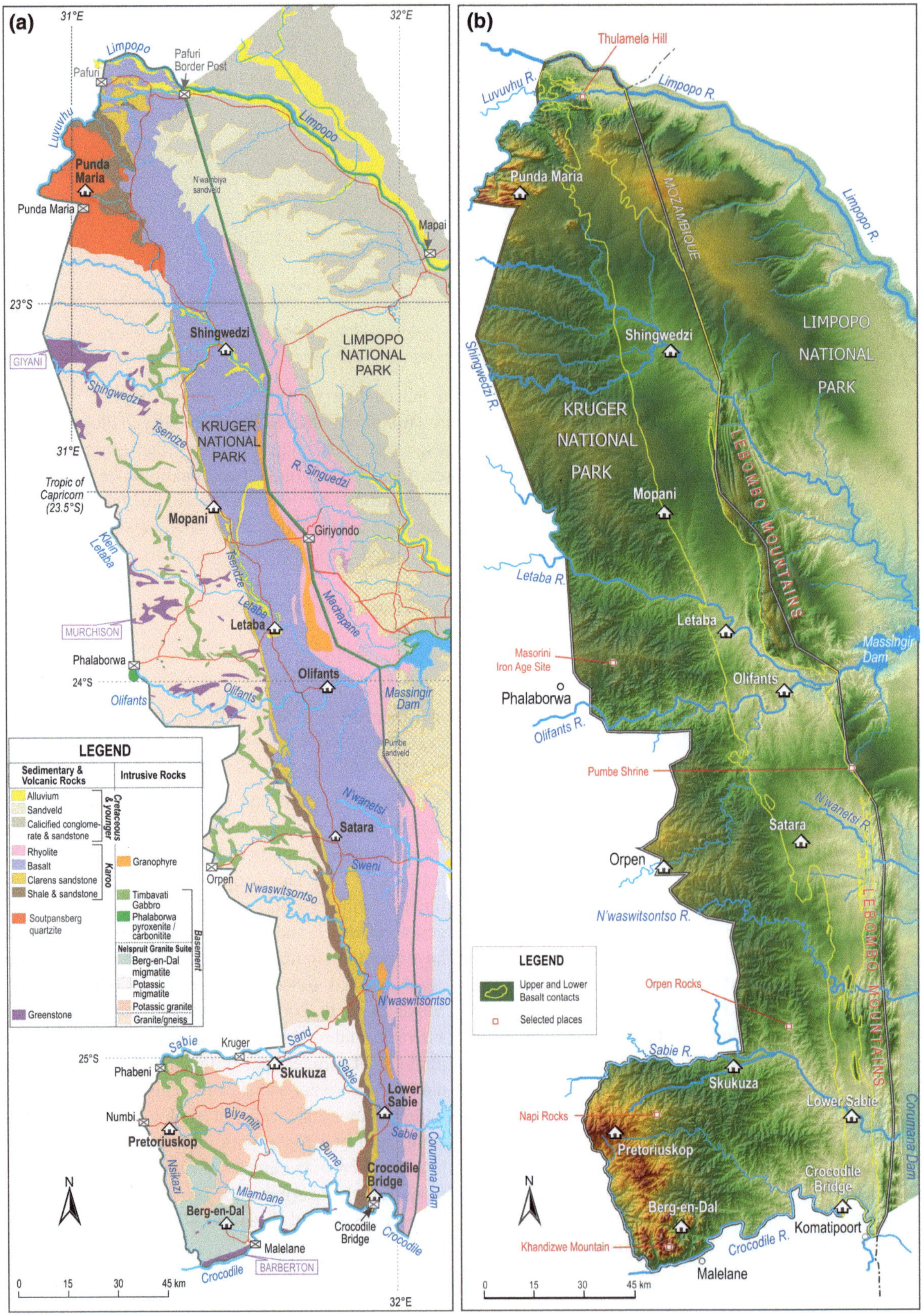

Fig. 13.1 **a** Geological map of Kruger and the western sector of the Transfrontier Limpopo National Park; **b** digital terrain image covering the same area. Comparison of the two highlights the relationship between geology and topography

13.2 Geological Setting

The general geologic setting of Kruger has previously been summarised by Bristow et al. (1986) and Schutte (1986). The western sector of Kruger is underlain by an ancient granite and greenstone basement dated to ~3.6–2.7 Ga. The southern boundary of Kruger (defined by the Crocodile River) flanks the northern edge of the Barberton greenstone belt, while the eastern extremities of the Murchison and Giyani greenstone belts terminate in the western part of central and northern Kruger (Fig. 13.1). The far north-western sector of Kruger is underlain in part by red sandstones and conglomerates of the Soutpansberg Group (~1.9 Ga in age), which farther west build the main Soutpansberg mountain range.

Sedimentary and volcanic rocks of the Karoo Supergroup underlie the entire eastern sector of Kruger. They dip at shallow angles to the east and straddle the Mozambique border where resistant felsic volcanic rocks form the Lebombo Mountains (Fig. 13.1). Shales and sandstones of the lower Karoo sequence overlie basement rocks and were deposited from about 300 Ma. They occur as north–south trending belts throughout much of Kruger, but these rocks rarely outcrop on the surface. In the far north of the region, they host coal deposits and the earliest dinosaur fossil (the prosauropod *Massospondylus*). The Clarens Formation constitutes the uppermost sedimentary formation of the Karoo Supergroup and is prominent in far northern Kruger where it is preserved as a number of fault blocks of light yellowish brown sandstone. To the south, the Clarens Formation forms a narrow, sporadically outcropping belt of light yellow to orange coloured sandstone. Volcanic activity commenced at about 182 Ma, building a significant succession of gently east-dipping basaltic lavas that now form an ~20 km wide and 360 km long, poorly exposed, north–south trending belt in eastern Kruger (Fig. 13.1). Following this period, explosive eruptions of felsic and acid volcanics (mainly rhyolites) took place, and together with sub-volcanic intrusive rocks known as granophyres now form the Lebombo Mountains.

Cretaceous age (140–65 Ma) calcified gravels and sandstones (Malonga Formation) are extensively developed and exposed in the valleys of the Limpopo, Shingwedzi and Olifants rivers, mainly to the east of the Lebombo Mountains, in the Mozambique sector of the Transfrontier Park (Fig. 13.1). Relict gravel terrace remnants and extensive alluvial flats are features of the major river valleys of northern Kruger.

The varied underlying geology has played a fundamental role in developing the distinctive topographies, landscapes and landforms of Kruger (Fig. 13.1), which reflect the effects of climate, resistance to weathering of different rock types and erosion by major rivers (Venter and Bristow 1986). The geology has also influenced the development of soil, vegetation types and ultimately ecosystems, with an inextricable link between the abiotic and biotic (the 'Earth/life link'), which is better revealed in Kruger than perhaps anywhere else in South Africa (McCarthy and Rubidge 2005).

13.3 Landscapes and Landforms of Kruger

During the Cenozoic, Kruger underwent several episodes of uplift, planation and fluvial incision. Land surface processes during these episodes have been strongly influenced by different rock types of Kruger, which have given rise to distinctive geomorphic provinces, as described by Partridge et al. (2010). The Lowveld province developed through extensive erosion between the massively uplifted, resistant rocks of the eastern escarpment (Chap. 3) located to the west, and the slightly uplifted gently east-dipping rocks of the Lebombo Range to the east (McCarthy and Rubidge 2005). Other than for occasional granitic koppies (hillocks) in the south of Kruger, the remainder of the landscape is covered by extensive pediments, developed mostly during the post-Africa 1 erosional cycle around 20 Ma. The lowest lying basaltic lavas (Lebombo flats) immediately west of the Lebombo Highlands represent the post-African 2 surface, formed by tectonic uplift around 5 Ma (Partridge and Maud 1987; McCarthy and Rubidge 2005).

13.3.1 Landforms Associated with Rock Weathering

Pressure release joints that are formed within granites by the unloading of overburden can aid in the development of exfoliation slabs and spheroidal weathered forms (Migoń 2009). Exfoliation slabs that initially have angular edges weather and erode to eventually form spheroidal, boulder-strewn domes and monoliths. The resistant, homogeneous, more potassic and silica-rich granites of southern Kruger in particular give rise to large exfoliation domes, inselbergs and koppies (Fig. 13.2a). In some cases, remnant spheroidal boulders remain perched on the top of monoliths (Fig. 13.2b). Where diabase and dolerite dykes are present, and where Timbavati gabbro has intruded into the granitic rocks, typical spheroidal forms with the classic 'onion skin' morphology are a common outcome. The resistant upper part of the Timbavati gabbro sheet has formed jointed, blocky ridges such as Ship Hill in southern Kruger, Shilawari Hill south-west of Letaba camp, and Tshanga lookout south-west of Shingwedzi camp (Fig. 13.2c).

Quartz in sandstones is highly resistant to chemical weathering, so chemical decomposition of sandstone consists largely of an attack on the cementing materials (mainly calcite) or clay-rich layers within the sandstone. The relatively

Fig. 13.2 **a** Exfoliation dome of Shabeni Hill (759 m) and related granite domes located north of Pretoriuskop camp; **b** large monolithic granite boulder topped by two small spheroidal boulders representing the last remnants of an earlier exfoliation layer, Napi Boulder (505 m) with Ludick memorial plaque, located east of Pretoriuskop; **c** jointed blocky outcrop of resistant upper part of east-dipping Timbavati gabbro sheet, Tshanga lookout, located south-west of Shingwedzi camp; **d** cavernous weathering of Clarens sandstone, west of Crocodile Bridge camp (*Photographs* M. Viljoen)

massive Clarens Formation sandstone forms ridges and low, domical outcrops in northern Kruger. Cavernous weathering of clay-rich sandstone units forms honeycombs and larger cavernous hollows (tafoni), often capped by more resistant sandy overhangs (Fig. 13.2d). Basalt, another dominant rock type in Kruger, is fine grained and composed mostly of feldspar, pyroxene and olivine. Chemical weathering readily reduces these minerals to clays and iron oxides, the final weathering product of which is a red or brown soil which now covers flat, featureless plains of eastern Kruger (Fig. 13.1). By contrast, rhyolitic flows, together with intrusive granophyre dykes and sills, are very resistant to weathering due to their high silica content. North–south trending granophyre dykes, west of Olifants camp, form distinctive, low and narrow spheroidally weathered/eroded ridges.

13.3.2 Erosional Fluvial Landforms

Different fluvial landscape features have developed on the varied rock types of Kruger. The dominant granitic rocks of western Kruger host a dense dendritic drainage pattern characterised by numerous mini watersheds and valleys, which serve as tributaries for major west–east rivers flowing across Kruger (Fig. 13.1). Drainage density on sedimentary and volcanic rocks elsewhere in Kruger is much lower. On the flat lava plains, isolated streams tend to form broad, rectangular patterns such as the Nwanetsi River, east of Satara camp (Venter and Bristow 1986) (Fig. 13.1). River incision through the resistant rhyolite lava ridges of the Lebombo Mountains has given rise to a variety of fluvial erosional features, the most spectacular being the gorges of the Shingwedzi, Olifants, Nwaswitsontso and Sabie rivers (Figs. 13.1, 13.3 and 13.4a).

Rivers that traverse the Lebombo Mountains are unique in that they flow orthogonally in narrow, deeply incised gorges (Venter and Bristow 1986; Partridge et al. 2010). This unusual hydrography reflects a former continuous pediment surface (the African surface) between the crest of the Lebombo in the east and the eastern escarpment to the west, which the rivers traversed. Rounded gravels containing clasts of hard quartzite from the eastern escarpment attest to the existence of such a drainage network at a level of ~400–800 m prior to the cycle of erosion which produced the Lowveld (Partridge et al. 2010). This orthogonal fluvial system transecting of the Lebombo thus reflects a superimposed drainage pattern (King 1963). The 840-m-high Khandizwe granitic peak near Malelane is an additional remnant of the former erosion surface (Du Toit 1954; Venter and Bristow 1986).

On the rhyolite bedrock, the rivers have exploited less resistant lava flow contacts as well as joints and fractures, aided in part by potholing, to produce linear channel reaches in different flow directions, resulting in a trellis fluvial drainage network (Fig. 13.4b). In part, this pattern grades into bedrock-influenced anastomosing channel types (Venter and Bristow 1986) (Fig. 13.4a, b).

A range of pothole structures are present wherever the main rivers traverse massive, homogeneous rock formations. Individual potholes are mostly circular and form mainly during floods when turbulent water drives hard pebbles and boulders along the river bed. In some instances, the circular grinding action enables these pebbles to bore downwards into underlying rocks. An outstanding example is the 'Red Rocks' site, where the Shingwedzi River has eroded a spectacular and extensive potholed surface in what is otherwise a dominant alluvial channel (Fig. 13.5). In some instances where potholes have coalesced, mini gorge-like incisions have developed. Mini gorges have also formed where the Olifants and Letaba rivers traverse basalts. In such instances, the mini gorges have formed almost entirely by the coalescing of potholes. The apexes of the mini gorges have migrated upstream to knick points that are frequently defined by waterfalls. Smaller and more abundant potholes have been preferentially eroded in the slightly less resistant upper parts of lava flows, with larger potholes dominating in the more massive lower parts of the individual flows. The grinding

Fig. 13.3 Resistant, light coloured rhyolite ridge overlying dark coloured basalt on the western flank of the Lebombo range. Junction of the Olifants and Letaba rivers (*Photograph* M. Viljoen)

(a)

(b)

Fig. 13.4 **a** The Olifants River gorge cutting through a resistant rhyolite of the Lebombo range close to the Mozambique border, and displaying trellis and bedrock-influenced anastomosing-type channel network. **b** Trellis and bedrock-influenced anastomosing-type channel network on the bed of the Olifants River where it cuts across the resistant rhyolite dome on which the Olifants camp is situated (*Photographs* M. Viljoen)

Fig. 13.5 Potholes cut into pavement of Clarens sandstone on the Shingwedzi River at Red Rocks (*Photograph* M. Viljoen)

stones found at the base of most potholes consist primarily of granite and pegmatite from the underlying basement.

The central and more southerly Kruger rivers (Olifants, Sand, Sabie, Crocodile) are geomorphologically most diverse in channel form, particularly on granitic bedrock, and include incised bedrock-influenced anabranching channel types, mixed anastomosing and pool-riffle channel types to fully braided alluvial channels (Van Coller et al. 1997; Heritage et al. 1999; Van Niekerk et al. 1999). In contrast, the more northerly rivers (Letaba, Shingwedzi, Luvhuvu, Limpopo) are predominantly alluvial in channel form, particularly where they traverse the flat basaltic lava plains (Fig. 13.1). All the major and secondary rivers traversing the Kruger are considered semi-arid (Rountree et al. 2000), with highly variable seasonal flow regimes, and some even cease flowing during most (e.g. Shingwedzi) or some (e.g. Letaba) of the dry seasons. The channel forms are under constant change, altered by deposition along the thalweg during low flow periods, and flushing of sediments during flood events. However, patterns of channel aggradation and erosion are also spatially variable and depend on bed slope and channel morphology (Parsons et al. 2006). These rivers experience prolonged (sometimes several decades duration) sediment accumulation phases, which are terminated or 'reset' by infrequent high-magnitude floods, as was the case for the southern Kruger rivers (Crocodile, Sabie, Olifants, Letaba) in 2000 and northern rivers (Shingwedzi, Luvhuvu) in 2013.

13.3.3 Depositional Fluvial Landforms

Recent alluvium is extensive along the Limpopo and Luvhuvu river floodplains of northern Kruger. In addition to the spectacular riparian vegetation, pans are also prominent features of these floodplains. The region contains 31 seasonally flooded pans, many of which may hold water long into the dry season and are surrounded by riverine forest, thereby creating important wildlife refuges (including for migratory water birds) in this otherwise dry landscape. The specific location of pans is strongly controlled by isolated outcrops of Clarens sandstone, basalt or older cemented gravels which create surface water gradients, leading to scouring and pan formation. Levee deposits and older flood-channel edges have also created sites for pan formation. The fine overbank clay deposits have formed a largely impermeable base to many pans, resulting in their retention of water throughout much of the dry season. Overbank flood deposits are also found associated with several of the larger Kruger river systems. Within a few years of deposition, these develop into a sparsely vegetated hummocky topography.

13.4 Geologic and Geomorphic Controls on Soils and Vegetation

Underlying geology has strongly controlled the development of different soil and vegetation types in Kruger (Venter 1986; Khomo et al. 2011). Sandy soils with an abundance of free quartz characterise granitic areas of western Kruger. These sandy soils are deeper on higher ground and in higher rainfall areas (e.g. south-west Kruger) where they support moderately dense tree savannah of mixed Bushwillow woodlands. Very shallow sandy soils are found in the broader river valleys such as the Sabie, Crocodile, Olifants and Shingwedzi.

Very shallow and stony soils are developed in various parts of Kruger and on a range of different rock types on steep, hilly or incised areas with significant rock outcrops. In the Berg-en-Dal area of south-west Kruger, these soils support trees such as rock figs. Similar stony and rocky soils along the Lebombo Mountains give rise to Red Bushwillow mountainveld. The Clarens Formation sandstone hills of the Pafuri area are also covered by sandy soils where a distinctive Ridge Bushveld vegetation assemblage characterises the sandy mini ridges and plateaus.

Deep clayey soils are developed on the Karoo basalts and parts of the Timbavati gabbro of southern Kruger, supporting a more open tree savannah than the adjoining denser vegetated areas. Sodic duplex soils (high in sodium or calcium salts) typically form on the Karoo shales and sandstones of southern Kruger. Soils are poorly drained and support the formation of small pans.

13.5 Recent Geomorphology of Kruger and the Natural/Human Environment

13.5.1 Rivers and Their Associated Ecosystems

The relationships between geology, geomorphology and ecosystems have been examined in a number of Kruger studies (e.g. Gertenbach 1983; Munyati et al. 2013). The dramatically changing landscapes along Kruger rivers have resulted from the flow regime becoming increasingly variable over time, given possible climate change and increased water extraction and land cover changes in their upper catchments. Previously perennially and seasonally flowing rivers have in some cases become seasonal (e.g. Limpopo) and episodic (e.g. Shingwedzi) in flow regimes. When these rivers are active, however, the sediment loads are exceptionally large due to the highly erodible soils within the catchments. Recent major floods have dramatically altered the landscape by widening alluvial channels and destroying much of the former tree-lined river banks

Fig. 13.6 Thin exfoliation layer of Clarens sandstone with petroglyphs being destroyed by weathering; petroglyph site located north-east of Mopani Camp (*Photograph* M. Viljoen)

of the Shingwedzi and Olifants rivers. Fauna dependent on the dense riparian vegetation (e.g. Bushbuck, Nyala, Scops owl, Vultures) have consequently been detrimentally affected by such changes. Nevertheless, seasonal pans are now developed in newly formed deep alluvial scours along the Shingwedzi in response to the February 2013 flood, providing water to wildlife during the drier months. Larger channel deposits are generally unaffected by most annual summer flows, but smaller channels are more dynamically shaped by near-annual return period flows, in particular in central and southern Kruger rivers (Heritage et al. 2001).

13.5.2 Archaeology and Cultural Landscapes

The Kruger area preserves evidence of human occupation spanning around the last 1 Ma, from the Early Stone Age to present (Verhoef 1986). Over 255 archaeological sites have been recorded in the park. The distribution and to some extent the preservation of virtually all archaeological and historical sites in Kruger are controlled by geologic and landscape features described in this chapter. Early settlements were built on resistant hills and ridges largely for strategic reasons. An example is Thulamela Hill, a stone-walled citadel occupied from the fifteenth to the mid-seventeenth centuries and built on a Clarens Formation sandstone fault block overlooking the Luvuvhu River. A petroglyph site consisting of circular engravings, a cheetah and other features, is found on an exfoliation Clarens pavement, north-east of Mopani Camp (Fig. 13.6). Immediately south-east of the petroglyph site, ancient walls are preserved on a resistant basalt ridge. The Pumbe shrine, similar to Hindu places of worship in south India, is situated on the elevated sandveld overlying the Lebombo Mountains of south/central Kruger, and the Masorini Iron Age smelting site is situated on a resistant syenite intrusion that forms part of the 1,946 Ma Phalaborwa igneous intrusion.

Many of the historical commemorative plaques of southern Kruger are affixed to exfoliated and spheroidally weathered monoliths and boulders of the Nelspruit Granite Suite. They include Stevenson Hamilton, Kruger tablets, Orpen rocks and the Ludick memorial at Napi Rocks. Preservation of archaeological heritage and, in turn, cultural landscapes is as yet not fully integrated with ecosystem management plans. This is an important aspect for future landscape-scale management.

13.6 Conclusions

A number of varied geological formations in Kruger, together with climate variations, have played a fundamental role in determining its landscapes, soil and ecozones. The present geomorphic landscapes of Kruger rivers are experiencing environmental stress and (hydro-) morphological change, which is also driving ongoing ecological change. Park management has built dams across many of the rivers in the hope of providing more perennial water resources, but these have consequently impacted downstream hydrology, fluvial geomorphology and biodiversity (e.g. Power et al. 1996).

Many plant and animal species are endemic to particular landscape components in Kruger. For example, the Klipspringer antelope is almost exclusively found in the rocky, inselberg landscape, whereas rhino is often seen on the ecozones associated with the darker soils of the Timbavati gabbro, particularly in southern Kruger. The link between the abiotic and biotic (Earth-Life Link) is perhaps better illustrated in Kruger than almost anywhere else in South Africa, and the interpretation of this phenomenon for education, or to enhance the tourism experience in Kruger, should be an important ongoing goal.

Acknowledgments The input of MV's co-authors on the book: *Earth/Life Link in the Kruger Park* (in prep.), Vanessa Strydom and Ingrid Booysen, is gratefully acknowledged. Ingrid Booysen is thanked for drafting the map. Improvements to the manuscript, offered by two anonymous referees, Stefan Grab, Jasper Knight, Richard Viljoen and Vanessa Strydom, are gratefully acknowledged.

References

Bristow JW, Sweeney R, Venter FJ (1986) Geological guide to selected areas of the Kruger National Park. Koedoe 29:183–202

Du Toit AL (1954) The geology of South Africa, 3rd edn. Oliver and Boyd, Edinburgh

Gertenbach WPD (1983) Landscapes of the Kruger National Park. Koedoe 26:9–122

Heritage GL, van Niekerk AW, Moon BP (1999) Geomorphology of the Sabie River, South Africa: an incised, bedrock-influenced channel. In: Miller AJ, Gupta A (eds) Varieties of fluvial form. Wiley, Chichester, 53–79

Heritage GL, Moon BP, Jewitt GP, Large ARG, Rountree M (2001) The February 2000 floods on the Sabie River, South Africa: an examination of their magnitude and frequency. Koedoe 44:37–44

King LC (1963) South African Scenery (3rd edn), Oliver and Boyd, London, 308 pp

Khomo L, Hartshorn AS, Rogers KH, Chadwick OA (2011) Impact of rainfall and topography on the distribution of clays and major cations in granitic catenas of southern Africa. Catena 87:119–128

McCarthy TS, Rubidge B (2005) The story of Earth and Life: a southern Africa perspective of a 4.6 billion year journey. Struik Publishers, Johannesburg

Migoń P (2009) Are any granite landscapes distinctive of the humid tropics? Reconsidering multiconvex topographies. Singap J Trop Geogr 30:327–342

Munyati C, Ratshibvumo T, Ogola J (2013) Landsat TM image segmentation for delineating geological zone correlated vegetation stratification in the Kruger National Park, South Africa. Phys Chem Earth 55–57:1–10

Mutanga O, Prins HHT, Skidmore AK, Wieren S, Huizing H, Grant R, Peel M, Biggs H (2004) Explaining grass-nutrient patterns in a savanna rangeland of southern Africa. J Biogeog 31:819–829

Parsons M, McLoughlin CA, Rountree MW, Rogers KH (2006) The biotic and abiotic legacy of a large infrequent flood disturbance in the Sabie River, South Africa. River Res Appl 22:187–201

Partridge TC, Maud RR (1987) Geomorphic evolution of southern Africa since the Mesozoic. S Afr J Geol 90:179–208

Partridge TC, Dollar ESJ, Moolman J, Dollar LH (2010) The geomorphic provinces of South Africa, Lesotho and Swaziland: a physiographic subdivision for earth and environmental scientists. Trans R Soc South Africa 65:1–47

Power ME, Dietrich WE, Finlay JC (1996) Dams and downstream aquatic biodiversity: potential food web consequences of hydrologic and geomorphic change. Environ Manage 20:887–895

Rountree MW, Rogers KH, Heritage GL (2000) Landscape state change in the semi-arid Sabie River, Kruger National Park, in response to flood and drought. S Afr Geogr J 82:173–181

Schutte IC (1986) The general geology of the Kruger National Park. Koedoe 29:13–37

Van Coller AL, Rogers KH, Heritage GL (1997) Linking riparian vegetation types and fluvial geomorphology along the Sabie River within the Kruger National Park, South Africa. Afr J Ecol 35:194–212

Van Niekerk AW, Heritage GL, Broadhurst LW, Moon BP (1999) Bedrock anastomosing channel systems: morphology and dynamics of the Sabie River, Mpumalanga Province, South Africa. In: Miller P, Gupta AJ (eds) Varieties of fluvial form. Wiley, Chichester, pp 33–51

Venter FJ (1986) Soil patterns associated with the major geological units of the Kruger National Park. Koedoe 29:125–138

Venter FJ, Bristow JW (1986) An account of the geomorphology and drainage of the Kruger National Park. Koedoe 29:117–124

Verhoef J (1986) Notes on archaeology and prehistoric mining in the Kruger National Park. Koedoe 29:149–156

The Maputaland Corridor: A Coastal Geomorphological Treasure

14

Greg Botha

Abstract

Maputaland represents the southern limit of the east African coastal plain, flanked by the volcanic Lebombo mountain cuesta and the high-energy Indian Ocean. The terrain morphology of the region bears testimony to deep erosional incision and dramatic sea-level fluctuations since the early Cretaceous. The topography of the coastal plain is closely linked to the sequence of aeolian sand deposits that have been differentially weathered and eroded. The late Neogene shoreline has been exploited by the Phongola River. Under the Tshongwe–Sihangwane sand megaridge and the wetland systems towards the east, the truncated dune landscape of the Kosi Bay Formation perches groundwater, providing groundwater seepage and seasonal run-off to the coastal lakes. The extended parabolic dune systems of the KwaMbonambi Formation define the surface relief and confine interdune wetlands. The coastal lakes have evolved in response to Pleistocene sea-level fluctuations that have inundated river valleys up to 25 km inland and caused foredunes to be submerged several kilometres offshore on the shelf. Accretion of the high coastal barrier dune cordon during the Holocene isolated some coastal lakes and forced morphological changes in others. Management of this area with international conservation status must draw from the dramatic influences that geomorphic processes have had on this region.

Keywords

iSimangaliso • Maputaland • Wetlands • Dunes • Lebombo

14.1 Introduction

The name Maputaland, referring to the area extending from Maputo Bay in Mozambique, southwards to St Lucia estuary, evolved from the subdivision of this southern part of the east African coastal plain by colonial powers and is home to the Thonga people (Bruton 1980). Although the name is still in use today, the area has been subdivided into municipalities and is marketed as an ecotourist destination: The Elephant Coast. The conservation value of the region resulted in the inscription of iSimangaliso Wetland Park in 1999 as South Africa's first UNESCO World Heritage Site, linking Lake St Lucia, Africa's largest estuary, with five ecosystems (http://www.isimangaliso.com/) and four Ramsar wetland sites. The 332,000-ha park includes 220 km of beaches and extends onto the continental shelf where the southernmost Indian Ocean coral reefs have colonized ancient submerged dunes. This geographically diverse region comprises a mosaic of landforms and habitats that attest to active fluvial, lacustrine, estuarine, high-energy beach and coastal dune processes (http://whc.unesco.org/en/list/914). The international conservation status of the region focuses research, international ecotourism and management priorities in the heterogeneous landscape, demanding an integrated, multi-disciplinary understanding of the landforms, geomorphic processes and biota to ensure that this region is conserved for future generations.

G. Botha (✉)
Council for Geoscience, Pietermaritzburg, South Africa
e-mail: gabotha@geoscience.org.za

S. Grab and J. Knight (eds.), *Landscapes and Landforms of South Africa*,
World Geomorphological Landscapes, DOI 10.1007/978-3-319-03560-4_14,

14.2 Geographical/Environmental Setting

The dramatic landscape of the south-eastern African coastal margin owes its origin to over 180 million years of substantially different geological and geomorphological processes. The last vestiges of the supercontinent—Gondwana—are preserved as the Lebombo Mountains: the volcanic rocks plunging deep below the coastal plain towards the zone of rifting that separated Africa from East Antarctica. The formation of the proto-Indian Ocean and its deepening as the continents separated is reflected in the thick succession of Cretaceous marine deposits that accumulated on the continental margin. These fine clastic sedimentary rocks contrast with the overlying Eocene marine carbonate rocks and Miocene to Holocene siliceous littoral deposits that owe their origin to sea-level fluctuations over the past ~12 Ma, when aeolian dune systems spread across the continental margin in response to sea level and cyclical climate changes (Botha et al. 2013). It is the transitional location between subtropical and tropical Africa that associates the region with the globally significant biodiversity of the Maputaland Centre of Endemism (Mathews et al. 2001), threatened ecosystems of the Maputaland–Pondoland–Albany Biodiversity Hotspot, and wetlands defined by the distribution of Freshwater Ecosystem Priority Areas (Nel et al. 2011). The close association between geology, landforms and vegetation types can be seen in Fig. 14.1.

The geological evolution of the Zululand and Maputaland Groups below the coastal plain has been described by Porat and Botha (2008) and Botha et al. (2013). The development of the Kosi lake system, Lake Sibaya and St Lucia has been described by Wright et al. (2000), Miller (2001) and Botha et al. (2013). The long period of post-emergence exposure and periods of stability between aeolian dune mobility events, linked to global climatic cycles, has resulted in differential weathering of the various bedrock types and unconsolidated Quaternary sediments that form distinct landscapes. Different degrees of pedogenesis over different timescales can be used to distinguish similar dune deposits of different ages (Botha and Porat 2007).

14.3 Evolutionary Process of Maputaland Landforms/Landscapes

14.3.1 Lebombo Mountains

The prominent Lebombo mountain cuesta rises up to 650 m a.s.l. The rugged rhyolite lava mountain range extends north of Mhlosinga for over 750 km, flanking the western boundary of the Maputaland and Mozambique coastal plain. The volcanic Lebombo Group comprises an estimated 6,500-m-thick basalt succession overlain by ~2,000 m of rhyolitic flows that form the homoclinal dip slope. Numerous streams have carved narrow gorges into the jointed rock mass, the most spectacular being the Phongola River gorge, dammed by the Pongolapoort Dam at Jozini, and the Ngwavuma and Usuthu River gorges farther north. The explosive volcanism responsible for extrusion of the rhyolitic pyroclastic flows during the late Jurassic (~178 Ma ago) was not responsible for the actual continental break-up and final rifting of Gondwana that occurred some 40 Ma later (Watkeys 2006). On the basal Lebombo footslopes within the Mkhuze Game Reserve, the Bumbeni Complex exposed around Mpilo Hill records the clastic sedimentation and volcanic caldera formation during several events from 146 to 133 Ma. The terminal volcanic events of the ~50 Ma Lebombo Group volcanics formed a pronounced ridge extending north-east beneath the coastal plain, separating the Cretaceous marine sequences into distinct sub-basins. Differential weathering has produced low-gradient footslopes that extend east of the Lebombo foothills where the resilient siliceous volcanic rocks are mantled by soft Cretaceous marine siltstones that weather to form red and black clayey soils on the Makatini Flats.

14.3.2 Cretaceous Zululand Group Marine Rocks

Progressive rifting of Gondwana formed the eastern margin of the emergent African continent as half graben rift basins formed during rapid subsidence associated with strike-slip movement on the plate bounding the Agulhas–Falkland Fracture Zone, forming the Mozambique Channel (Broad et al. 2006). Geophysical seismic profiling and oil exploration drilling reveal a deeply buried, narrow, fault-bounded graben beneath the remarkably linear western shoreline of Lake St Lucia, infilled with ~820 m of sediment that probably spans the Jurassic–Cretaceous boundary. This early rift deposit was buried by about 1,000 m of marine sediment when the early Cretaceous sea transgressed onto the continental margin.

The Zululand Group succession comprises up to 1,500 m of fossiliferous siltstones but is concealed beneath Cenozoic cover sediments over most of the Maputaland coastal plain. Terrestrial conglomerate and fossiliferous marine rocks of the basal Makatini Formation trace the early Cretaceous (Barremian to Aptian) marine transgression onto the continental margin around ~127 Ma (McMillan 2003). A depositional hiatus across the Aptian/Albian boundary (~113 Ma) marks the transition to the overlying Mzinene Formation succession of richly fossiliferous glauconitic siltstones that contain a diverse biota (Kennedy and Klinger

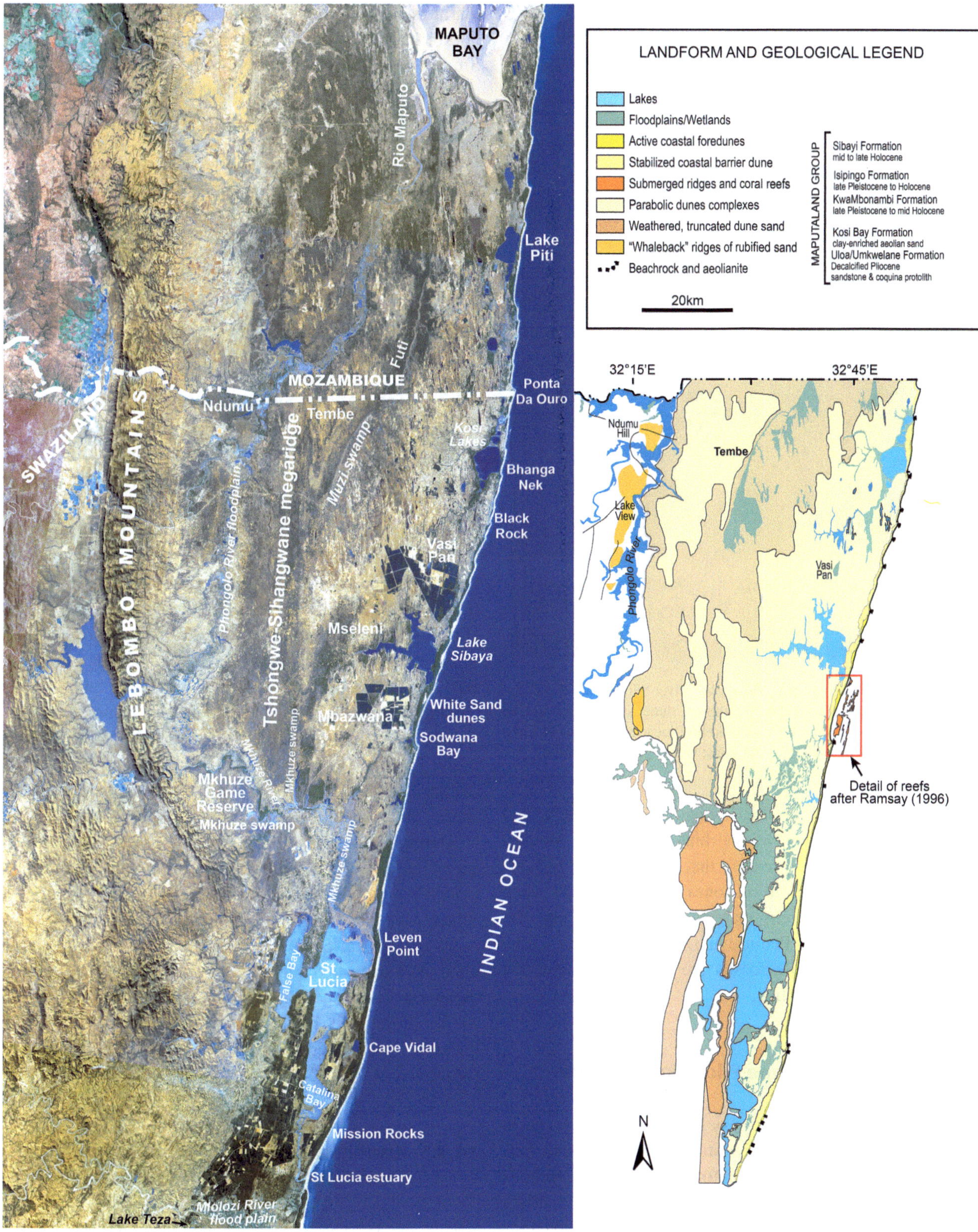

Fig. 14.1 Composite landsat image of greater Maputaland compared with a geological map of southern and central Maputaland. Landsat image courtesy of Dr B. Smith, University of Kent

Fig. 14.2 View across the Phongola River floodplain and the Maputaland coastal plain. The weathered dune sands (*red* in the foreground) lie above the Pliocene shoreline (near river level) at Lake View, now 60 km from the present coast (*Photograph* G. Botha)

1975), well exposed in the Phongola River valley and Ndumu Hill area. A depositional hiatus around 90 Ma ago marked the marine transgression that deposited the fossiliferous St Lucia Formation siltstones exposed in the steep cliffs defining the western margin of False Bay and the adjacent Nibela/Ndlosi Peninsula. During low lake levels, well preserved ammonite fossils are exposed around this section of the lake shoreline.

14.3.3 Uloa/Umkwelane Pliocene Shoreline

There is little evidence of the very high Eocene sea levels in Maputaland, except for the limestone exposed around Salamanga in the lower Rio Maputo valley. The Cretaceous rocks in Maputaland are generally unconformably overlain by Pliocene-age shallow marine deposits of the Uloa Formation and the overlying Umkwelane Formation aeolianite (Botha et al. 2013). A strandline at ~45 m a.s.l. has been exploited by the Phongola River downstream of Jozini where boulder beach conglomerates and shelly shallow marine sandstones containing oyster shells are exposed up to 60 km from the present Indian Ocean shoreline (Fig. 14.2). Similar deposits occur in the lower Msunduzi valley slopes and on the eastern side of Nibela peninsula in the lower Mkhuze valley. Weathered karstic remnants of this lithology underlie the coastal plain to the east where they represent an important groundwater aquifer. These deposits form the low 'whaleback' ridges below Mantuma Camp in Mkhuze Game Reserve and the western interfluve along the Phongola River valley from Mtoti pan to Lake View and Ndumu Hill. The ubiquitous residual red sand weathering product, known locally as 'Berea-type red sand', is underlain by a hard cemented ferruginous and silicified weathering profile.

14.3.4 Tshongwe–Sihangwane Megaridge

A 15-km-wide sand ridge extends from the Mkhuze valley near Tshongwe village to Sihangwane and Tembe Elephant Park close to the Mozambique border in the north, defining the interfluve with crest elevations at 90–125 m a.s.l. between the Phongola valley in the west, and the eastern Muzi swamps that drain the central part of Maputaland (Fig. 14.1). Water boreholes have intercepted the calcareous Uloa/Umkwelane Formation beneath the sand ridge. Much of the smooth central ridge was formed in sands of the Kosi Bay Formation which represent a weathered and truncated coastal plain dune landscape. Remnants of this eroded land surface have been revealed within the bed of Lake Nhlange in the Kosi lakes (Wright et al. 2000), around Lake Sibaya (Miller 2001) and at the Sibomvini ridge on the eastern shores of St Lucia.

During the period after the coldest, driest phase of the Last Glacial cycle, a narrow plume of KwaMbonambi Formation parabolic dunes extended north over the Tshongwe–Sihangwane megaridge from the Mkhuze valley between ~19 and 15 ka. Sand was deflated from the central ridge areas and deposited farther north as a complex parabolic dunefield that ramped onto the Sihangwane area from ~22–11 ka (Porat and Botha 2008). These sands support the endemic species-rich sand forest in Tembe Elephant Park (Matthews et al. 2001).

14.3.5 KwaMbonambi Formation Parabolic Dunes and Interdune Wetlands

East of the Tshongwe–Sihangwane megaridge, the complex wetland system feeding the Muzi–Futi channel dominates the central Maputaland coastal plain, draining north into the Rio Maputo (Fig. 14.1). The wetland complex is underlain by a truncated late Pleistocene dune landscape and thick

Fig. 14.3 Aerial view of mobile buttress and transverse foredunes along the base of the forested coastal barrier dune north of Cape Vidal with Lake Bhangazi (south) in the background (*Photograph* G. Botha)

alluvial clays with interbedded calcified diatomaceous deposits. Differential soil development in the dune sands permits the older, reddened and slightly clay-enriched, truncated Kosi Bay dune deposits to be distinguished from the younger, relatively unweathered KwaMbonambi Formation aeolian sand that forms the regional parabolic dune topography (Botha and Porat 2007). This mantle of remobilized aeolian sand extended north from the deflated Lake Sibaya basin as parallel elongated parabolic dunes, from 5 to 20 m high (Figs. 14.1 and 14.4). The truncated Kosi Bay Formation dune landscape was eroded and parabolic dunes were mobilized as the regional water table level dropped around 35–18 ka, in concert with marine regression leading up to the Last Glacial Maximum sea-level lowstand. Reduced vegetation cover resulted in deflation of exposed sand, forming blowouts from which parabolic dune noses migrated under the influence of prevailing southerly winds, leaving parallel trailing dune limbs bounding the deflated sand source area. Dune migration was controlled by the groundwater table and vegetation stabilization. The cores of some of the highest dune limbs date from ~57 to 22 ka but the majority date from ~15 to 7 ka (Porat and Botha 2008).

Interdune hollows were periodically flooded by a rising groundwater table to form open, freshwater lakes where diatomite accumulated. Deposits in the Muzi wetlands and on the interfluves around Mbazwana attest to periods of wetness after ~53 ka that were terminated by parabolic dune mobility during subsequent aridity after ~15 ka. A significant indicator of the effect of climate change on the coastal plain during the Holocene is the coincidence of cessation of aeolian sand mobilization around 7 ka with the accumulation of peat within the interdune wetlands such as Vasi Pan (Grundlingh et al. 1998). The Mfabeni swamp on the eastern shores of Lake St Lucia occupies a depression in the core of a former island which was morphologically similar to Bazaruto in Mozambique during the Last Interglacial. Peat formation began much earlier here, around 45 ka, and continued through the Holocene (Finch and Hill 2008). Stream incision related to lowering of sea level can be used as a relative dating tool to distinguish the older, pre-18-ka KwaMbonambi Formation from the unincised terminal Pleistocene and Holocene dunes.

14.3.6 Forested Composite Coastal Barrier Dune

The forested coastal barrier dunes extending north from St Lucia include parabolic dunes rising to 172 m a.s.l and are often cited as some of the highest coastal dunes in the world (Figs. 14.3 and 14.5). However, this Sibayi Formation barrier dune has a complex history represented by four phases of Holocene sand accretion. The distinctive composite transverse ridge morphology is formed by closely spaced parabolic dunes which ramped onto a semi-consolidated core of calcified dunes that date to around 130 ka, with the ~400 ka Port Durnford Formation estuarine deposits buried beneath the Kosi Bay Formation.

Along much of the coast, successive parabolic dune accretion events were influenced by southerly winds. North of Cape Vidal, however, the youngest phase of barrier dune accretion was by large dunes that moved towards the south-

Fig. 14.4 View of eastward over Lake Sibaya showing the flooded parabolic dune topography outlined by the sinuous shoreline and the high coastal barrier dune cordon in the far distance that isolated the lake during the Holocene (*Photograph* G. Botha)

Fig. 14.5 Dumile dune (*upper right*) on the forested coastal barrier close to the southern embayment of Lake Sibaya. The White Sands dunefield (*upper left*) is stabilized by *Casuarina* plantations. In the foreground, a submerged beachrock formation defines a log-spiral embayment similar to those that occur along the present intertidal zone (*Photograph* G. Botha)

west, possibly blocking a former outlet of the north-eastern shallows of Lake St Lucia near Leven Point (Botha et al. 2013). Although largely stabilized by coastal forests, localized areas of sand mobility occur within interdune hollows. One of the largest mobile sand areas is the White Sands dunefield north of Sodwana Bay where wind-blown sand deflated from foredunes form buttress and transverse dunes that have migrated north over the barrier dune towards a southern basin of Lake Sibaya before being stabilized by *Casuarina* plantations (Fig. 14.5).

14.3.7 Polygenetic Coastal Lakes

The coastal lakes have a long history that is closely linked to cyclical glacioeustatic sea-level fluctuations and to periods of dune development. Sea level has been lower than present for much of the past 2 Ma, peaking for a period of a few thousand years during the Last Interglacial (Marine Isotope Stage 5e) around 125 ka. The Kosi lakes, Lake Sibaya (Fig. 14.4), St Lucia and smaller lakes such as lakes Bhangazi (north and south) and Mgobozeleni all share a similar history. Within False Bay and its northern tributary, the Mzinene River, fossiliferous marine conglomerates represent a period of raised sea level (+6 m a.s.l.) when, probably during the Last Interglacial highstand, the north lake embayment of St Lucia had a direct marine link, and corals suggest that the water must have had low turbidity. As with Sibaya and St Lucia, this marine link was progressively closed by coastal barrier dune accretion. Within the past 3 ka, the mouth of the Kosi system was forced to migrate northwards to its present position when barrier dunes closed the old estuary at Bhanga Nek (Wright et al. 2000). These lake basins were only inundated for short periods during the past 200 ka and for much of that time were desiccated coastal flats drained by shallow streams. During the marine transgression from the lowstand of −130 m about 18 ka (Green and Uken 2005), the lake embayments were gradually inundated. The morphology of Lake Sibaya is derived from inundation of the surrounding dunefield topography and backflooding of shallow valleys. Sediments infilling the buried former fluvial channels show how the sedimentation balance shifted from marine to lacustrine conditions within the past 6 ka. A sequence of sandy beach ridges surrounding parts of the St Lucia lakes shows how the marine-influenced water level declined after the mid-Holocene highstand, reaching present levels within the past 1,000 years (Botha et al. 2013). Wind-driven waves and longshore currents are responsible for the erosion of sandy lake shorelines and segmentation of the water body by spit growth. Sedimentary infill sequences within tributaries of the Mkhuze and Mfolozi rivers provide evidence of marine influences more than 25 km from the coast during the Holocene (Neumann et al. 2010).

14.3.8 Wetland Landscapes

The scenery of the iSimangaliso landscape is enhanced by the topographic contrast between high coastal barrier dunes and low-lying wetland systems feeding the coastal lakes. The role of wetlands in landscape development and in sustaining the lakes is intricately linked to the various ages of aeolian sand deposition. Lake Sibaya and some of the smaller wetlands such as Mgobozeleni and Bhangazi (north and south) have limited fluvial channel catchments and groundwater baseflow is an important contribution that offsets evaporation. Indeed, the resilience of Lake St Lucia biota to periods of high salinity and low lake levels is reliant on groundwater seepage (Taylor et al. 2006). The long-term evolution of interdune wetlands and vegetation has been derived from analysis of fossil pollen preserved in the peat, yielding radiocarbon dates pointing to accumulation of organic matter in these systems (Grundlingh et al. 1998; Finch and Hill 2008).

14.3.9 Coastline and Continental Shelf Landforms

The linearity of Maputaland's coastline owes its origin to the structural influence of beachrock within the intertidal zone, prevailing winds, longshore currents, and the lack of major river valleys north of the Mfolozi/St Lucia estuary. Sandy beaches are influenced by the energetic wave regime and longshore currents as well as the presence of several generations of carbonate cemented beachrock and cemented Pleistocene aeolian or beach facies. Aeolianite or beachrock forms the typical rocky headlands such as Jesser Point at Sodwana Bay. The excavation of narrow rocky intertidal platforms on the northern side of headlands by longshore currents forms characteristic log-spiral embayments. The large-scale planar cross-beds of the calcified Isipingo Formation aeolian dune deposits are preserved at Bat's Cave north of Mission Rocks and numerous other rocky points.

Fluctuating sea level is a central theme in the geomorphic development of the Maputaland coastal plain. Some might be surprised to find coast parallel dune ridges extending subparallel to the coastline on the narrow continental shelf (Ramsay 1996) where they were deposited as beach dune forms during periods of lower sea level (Fig. 14.1). The narrow continental shelf is swept by the south-flowing Agulhas Current which transports sand bedload as dune bedforms, depositing the sediment in canyons that cut the shelf slope. Much of the Maputaland shelf is sediment starved due to the lack of fluvial sediment input north of the Mfolozi River.

A characteristic feature of the Maputaland continental shelf is the occurrence of steep sided canyons incised into the shelf break slope that could represent fluvial channel landforms incised during the Last Glacial Maximum sea-level lowstand (Green 2009). The continued existence of the *Coelacanth* fish has been linked to these canyons; a specimen having been photographed in a canyon sidewall cave with a Last Glacial Maximum beach substrate.

14.4 Threats from Future Environmental Change

'iSimangaliso must be the only place on the globe where the oldest land mammal (the rhinoceros) and the world's biggest terrestrial mammal (the elephant) share an ecosystem with the world's oldest fish (the coelacanth) and the world's biggest marine mammal (the whale)' (Nelson Mandela; http://www.isimangaliso.com/). The management of such dynamic ecosystems in the face of global change that will affect regional rainfall patterns and sea level must take evidence of the effects of past environmental change into account. Removal of alien trees planted to stabilize mobile dunes on the coastal barrier will result in mobile sand encroachment onto coastal forest. The process of reversing the effects of physical separation of the Mfolozi River and St Lucia estuary mouths is underway to ensure increased freshwater inflow to the lake system. The impact of rising sea level could enhance sand accretion on foredunes, beach erosion and flood tide deposition in the estuary mouths. The pressure on groundwater in the catchments of Lake Sibaya could lower the lake level; widespread dune formation during recent times bears witness to the last period of lowered lake level. Many decisions that could be influenced on the basis of understanding the geomorphological processes that have shaped the Maputaland coastal plain will create management dilemmas. An example is whether the system is managed to minimize the impacts of gradual environmental change on the environment, or whether management practices adapt to these geomorphological processes.

14.5 Summary

The 180-Ma geological history of the Maputaland coastal plain has left landforms that witnessed the volcanic precursor to rifting of Gondwana and the influx of the proto-Indian Ocean waters onto the continental margin. Sea-level fluctuations have influenced the full width of the coastal plain within the past 4 Ma, and the coastline bears witness to shorter term cyclic climate change and sea-level fluctuations. This region has been granted international conservation significance in the light of globally significant endemism, biodiversity and scenic beauty. Management practices for conservation must adapt to the varied geology, landforms and geomorphic processes of this ancient landscape. The geological history bears witness to the dramatic environmental changes that occur rapidly in response to global climatic cycles, and our strategy to cope with these influences should consider lessons from the past.

Acknowledgements Research in Maputaland has benefitted from many years of support by the Council for Geoscience, from eZemvelo KZN Wildlife staff including Ricky Taylor and Caroline Fox and the iSimangaliso Wetland Park Authority. The composite Maputaland satellite image is provided courtesy of Dr Bob Smith of the Durrell Institute of Conservation and Ecology, School of Anthropology and Conservation, University of Kent, United Kingdom.

References

Botha GA, Porat N (2007) Soil chronosequence development on the southeast African coastal plain, Maputaland, South Africa. Quatern Int 162–163:111–132

Botha GA, Haldorsen S, Porat N (2013) Geological history. In: Perissonoto R, Stretch D, Taylor RH (eds) Ecology and conservation of estuarine ecosystems; Lake St Lucia as a global model. Cambridge University Press, Cambridge, pp 47–61

Broad DS, Jungslager EHA, McLachlan IR, Roux J (2006) Offshore mesozoic basins. In: Johnson MR, Anhaeusser CR, Thomas RJ (eds) The Geology of South Africa. The Geological Society of South Africa and Council for Geoscience, Pretoria, pp 553–571

Bruton MN (1980) A note on the name 'Maputaland'. In: Bruton MN, Cooper KH (eds) Studies on the ecology of Maputaland. Rhodes University and the Natal Branch of the Wildlife Society of South Africa, pp 533–536

Finch JM, Hill TR (2008) A late quaternary pollen sequence from Mfabeni Peatland, South Africa: reconstructing forest history in Maputaland. Quatern Int 70:442–450

Green AN (2009) Sediment dynamics on the narrow, canyon-incised and current-swept shelf of the northern KwaZulu-Natal continental shelf, South Africa. DOI, GeoMarine Lett. 29:201–219

Green AN, Uken R (2005) First observations of sea-level indicators related to glacial maxima at Sodwana Bay, northern KwaZulu-Natal. S Afr J Sci 101:236–238

Grundlingh P-L, Mazus H, Baartman L (1998) Peat resources in the northern KwaZulu-Natal wetlands: Maputaland. Department of Environmental Affairs and Tourism, Pretoria, 102 p

Kennedey WJ, Klinger HC (1975) Cretaceous faunas from Zululand and Natal, South Africa. Introduction, Stratigraphy. Bull Br Mus Nat Hist Geol 25:265–315

Matthews WS, Van Wyk AE, Van Rooyen N, Botha GA (2001) The vegetation of the Tembe Elephant Park, Maputaland, northern KwaZulu-Natal. S Afr J Bot 67:573–594

McMillan IK (2003) Foraminiferally defined biostratigraphic episodes and sedimentation pattern of the Cretaceous drift succession (Early Barremian to Late Maastrichtian) in seven basins on the South African and southern Namibian continental margin. S Afr J Sci 99:537–576

Miller WR (2001) The bathymetry, sedimentology and seismic stratigraphy of Lake Sibaya, northern KwaZulu-Natal, South Africa. Bulletin 131, Council for Geoscience, South Africa, 94 pp

Nel JL, Driver A, Strydom WF, Maherry A, Petersen C, Hill L, Roux DJ, Nienaber S, van Deventer H, Swartz E, Smith-Adao LB (2011) Atlas of freshwater ecosystem priority areas in South Africa: maps to support sustainable development of water resources. WRC report No. TT 500/11, 61 pp

Neumann F, Scott L, Bousman CB, van As L (2010) A Holocene pollen sequence and vegetation changes at Lake Eteza, KwaZulu-Natal (South Africa). Rev Palaeobot Palynol 162:39–53

Porat N, Botha GA (2008) The chronology of dune development on the Maputaland coastal plain, southeast Africa. Quatern Sci Rev 27:1024–1046

Ramsay PJ (1996) Quaternary marine geology of the Sodwana Bay continental shelf, northern KwaZulu-Natal. Bulletin, Council for Geoscience 117, 86 pp

Taylor R, Kelbe B, Haldorsen S, Botha GA, Været L, Wejden B, Bjørkenes MS (2006) Groundwater-dependent ecology of the shoreline of the subtropical Lake St Lucia estuary. Environ Geol 49:586–600

Watkeys MK (2006) Gondwana break-up: a South African perspective. In: Johnson MR, Anhaeusser CR, Thomas RJ (eds) The Geology of South Africa. The Geological Society of South Africa and Council for Geoscience, Pretoria, pp 531–539

Wright CI, Miller WR, Cooper JAG (2000) The late Cenozoic evolution of coastal water bodies in Northern KwaZulu-Natal, South Africa. Mar Geol 167:207–229

15 Of Dunes, Depressions and Dry Valleys: The Arid Landscapes of the Kalahari Desert

David J. Nash

Abstract

The combination of vegetated orange–red dunes, seasonal pans and dry valleys in the Kalahari creates a landscape with outstanding scientific and aesthetic value. This chapter describes the geomorphological features of the Kalahari Desert within South Africa and adjacent areas of Botswana and Namibia, with a special emphasis on aspects that make the landscape unique. The Kalahari is an arid to semi-arid region underlain by Cretaceous to recent Kalahari Group sediments, including a surface blanket of unconsolidated Kalahari sands. The landscape is dominated by three sets of landforms: (a) dry valley systems, including the Auob, Nossob, Kuruman and Molopo rivers; (b) partially vegetated linear dunes, which stretch in a broad zone from Upington on the Orange River into Botswana and Namibia; and (c) seasonally flooded pans. The importance of the long-term geological history of the Kalahari for understanding the present landscape is also discussed.

Keywords

Kalahari Desert • Dunes • Ephemeral drainage • Pan depressions • Duricrusts

15.1 Introduction

The northernmost Northern Cape and parts of North West Province in South Africa are situated within the Kalahari Desert, a vast inland sand sea at an elevation of ~1,000 m a.s.l. that covers over 2.5 million km^2 of South Africa, Namibia, Botswana, Angola, Zimbabwe, Zambia and the Democratic Republic of Congo (Thomas and Shaw 1991). The use of the term 'desert' to describe the Kalahari is rather misleading—whilst much of the area is covered by sandy sediment and is technically either arid or semi-arid (i.e. it has a mean rainfall of less than 500 mm per year), it is reasonably well vegetated with shrub or bush savanna. It is only really in its extreme south and south-west (Fig. 15.1), in an area straddling the Northern Cape, south-west Botswana and south-east Namibia, that the Kalahari takes on a true desert-like appearance. The landscape in this region, the focus of this chapter, is covered by sparse shrubs and grasses and contains a distinctive suite of landforms. These include extensive areas of partially vegetated dunes, seasonally inundated pan depressions and a network of dry valleys that occasionally flood following periods of heavy rainfall. The typical characteristics of the Kalahari landscape are well exhibited within the Kgalagadi Transfrontier Park which covers over 38,000 km^2 of south-west Botswana and adjacent areas of the Northern Cape.

The aim of this chapter is to summarise current understanding of the main components of the contemporary Kalahari landscape in southern Africa. It starts with a brief overview of the geological history of the Kalahari Basin since the Mesozoic, before moving on to consider the geomorphology of valleys, dunes and pans in the region. For simplicity, the Kalahari is defined here as encompassing all areas of South Africa and adjacent parts of Botswana and Namibia with an arid to semi-arid climate and that are underlain by Kalahari Group sediments (see Sect. 15.2).

D.J. Nash (✉)
University of Brighton, Brighton, UK
e-mail: d.j.nash@brighton.ac.uk

D.J. Nash
University of the Witwatersrand, Johannesburg, South Africa

S. Grab and J. Knight (eds.), *Landscapes and Landforms of South Africa*,
World Geomorphological Landscapes, DOI 10.1007/978-3-319-03560-4_15,

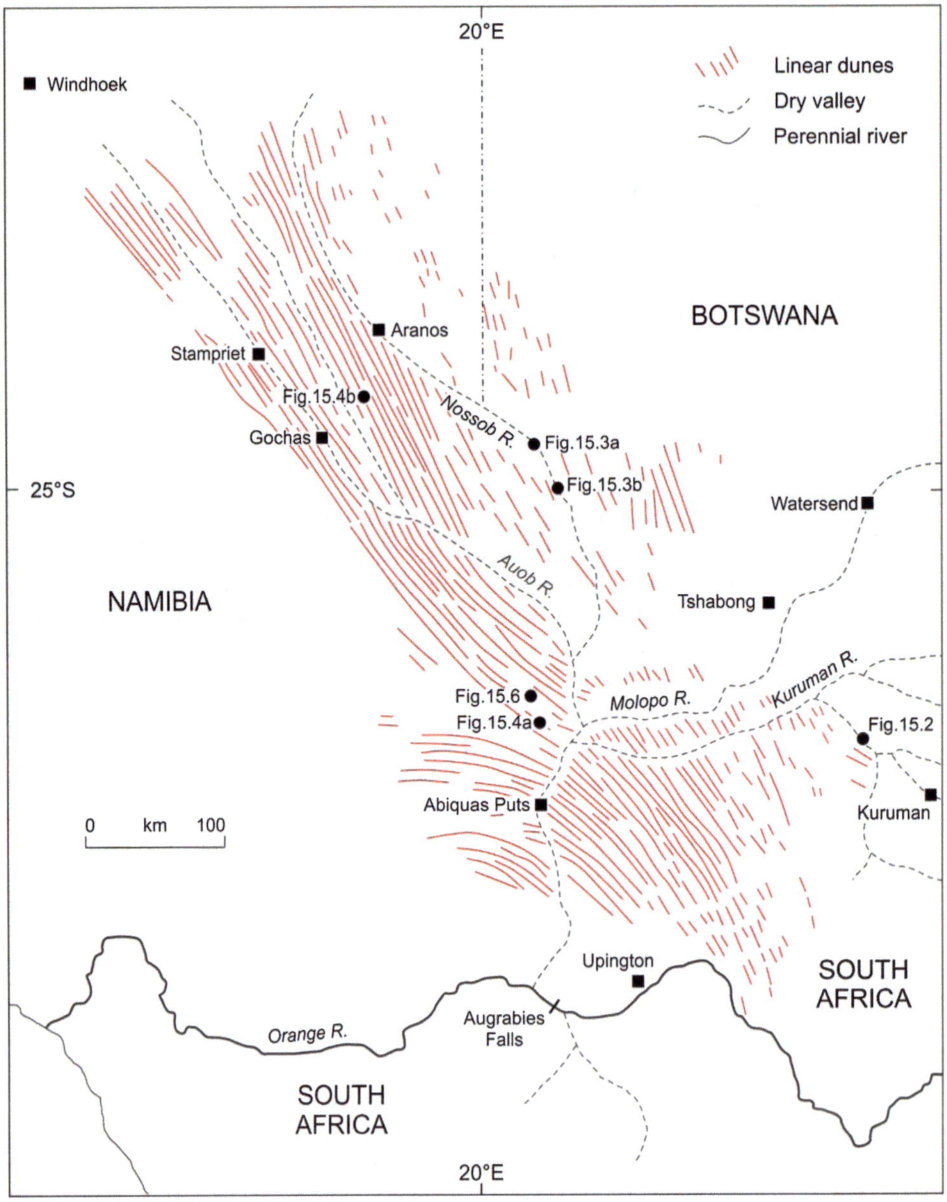

Fig. 15.1 Major geomorphological features of the south-west Kalahari Desert (modified after Bullard et al. 1995)

Readers wishing to know more about long-term landscape evolution are directed to the excellent overviews by Moore and Larkin (2001) and Haddon and McCarthy (2005). Evidence for Late Quaternary to historical environmental change is summarised in Thomas and Shaw (2002), Nash and Endfield (2002a) and Thomas and Burrough (2013).

15.2 Long-Term Landscape Evolution

The Kalahari landscape that we see today—essentially an inland sand sea sitting within an uplifted sedimentary basin—is the end product of a variety of interacting tectonic and geomorphological processes that have operated over considerable timescales. The initial development of the Kalahari was closely linked to the evolution of the African landmass following the break-up of the supercontinent Gondwanaland (Thomas and Shaw 1991) during the Mesozoic. The gradual separation of Africa from present-day Australia, Antarctica, India and South America was associated with progressive heating and rifting of the crust, which led to the uplift of the southern African continental margin in the mid- to Late Cretaceous. The development of this uplifted margin had significant implications for regional fluvial systems; the end product was a dual drainage pattern consisting primarily of short rivers draining directly into the sea off the Great Escarpment and longer rivers draining inland and never reaching the sea. Initial drainage patterns in southern Africa were dominated by several rivers which flowed south-eastwards from Angola before entering the Indian Ocean via the Limpopo, together with rivers draining westwards. Subsidence of the continental interior in the Late Cretaceous back-tilted sections of this drainage resulted in sediment deposition in the newly formed Kalahari Basin. These deposits—the Kalahari Group sediments—filled the Basin during the Late Cretaceous and Paleogene, with some deposition and subsequent reworking by the wind (Haddon and McCarthy 2005).

Additional episodes of tectonic uplift and downwarping are known to have occurred during the Paleogene, Neogene and Quaternary, mainly along distinct zones or axes (Moore and Larkin 2001). These had further impacts upon regional topography and drainage systems, including changes to the course of the Orange River. By the Miocene, the capture of the middle Orange by the lower Orange had already occurred, with the present course of the river established by the Late Oligocene. Many of the older components of the Kalahari Group are believed to have been cemented by silica and carbonate during a period of relative tectonic stability in the mid-Miocene to form thick sequences of silcrete and calcrete duricrusts (Haddon and McCarthy 2005). These crusts now exert a considerable influence on the landscape of the south-west Kalahari where they form an extensive calcrete-supported plateau surface. Further reworking by wind of the uppermost unconsolidated component of the Kalahari Group—the Kalahari sands—took place during the Pliocene and Quaternary (Thomas and Burrough 2013), with additional episodes of calcrete formation in the Late Quaternary (Nash and McLaren 2003). Most recently, human impacts, particularly to regional water tables and semi-arid vegetation systems, have occurred.

15.3 Dry Valleys

Despite being essentially a desert, dry valleys form a conspicuous part of the Kalahari landscape. These developed over much longer timescales than the current dunes discussed in Sect. 15.4. The main systems in South Africa are the north–south draining Auob and Nossob rivers, and the east–west draining Kuruman and Molopo (Fig. 15.1). These systems once connected to the Atlantic Ocean via a now-dry confluence of the Molopo and the Orange River downstream of Augrabies Falls (see Chap. 8). The four systems and their tributaries together form what Thomas and Shaw (1991) refer to as the southern Kalahari drainage; this is distinct from the predominantly fossil Middle Kalahari systems that once drained internally towards the Makgadikgadi depression or Okavango River and Delta in Botswana (see Nash 1996).

The cross-sectional shape of the Auob, Nossob, Kuruman and Molopo varies considerably, ranging from gently sloping convex/concave forms to steep-sided gorge-like sections where the valleys cut through the south-west Kalahari calcrete plateau (Nash et al. 1994) (Fig. 15.2). Variability occurs not only between the different systems but also along the lengths of individual valleys. The Auob valley, for example, is narrow and barely distinguishable in its headwater sections, but reaches a maximum width of 1.8 km south of Stampriet in Namibia before narrowing to around 0.5 km within the Kgalagadi Transfrontier Park.

All of the main southern Kalahari valley systems rise in areas of bedrock at or beyond the margin of the Kalahari Group sediments, although some tributary systems—such as the Moselebe valley, a feeder to the Molopo in Botswana—originate in areas with a Kalahari sand cover. As a result, although dry over much of their length, main valleys frequently contain water in their headwater sections. Headwater flows occur due to the effects of either heavier precipitation over higher elevation areas or groundwater discharge. The Auob and Nossob, for example, rise in the highlands of central Namibia. Both have a flashy regime, but flow regularly reaches as far as Aranos in the case of the Nossob (Leistner 1967) and Gochas for the Auob (Range 1912). The main Kuruman and Molopo valleys, in contrast, both originate at spring sites. The Kuruman has the most consistent flow of all southern Kalahari networks and is effectively a permanent river over the first 10 km of its course due to the 750 m^3 of water per hour supplied by its main spring source, the 'Eye' at Kuruman (Shaw et al. 1992). The Molopo only rarely contains water as far west as Bray on the Botswana border (Grove 1969), with the furthest recorded historical flow being at Watersend Farm (Shaw et al. 1992). Present-day flow conditions in all four systems have been greatly influenced by groundwater extraction. The waters of the Kuruman, for example, are used for irrigation and public supply, with the 'Eye' and other springs being exploited since at least the 1820s (Livingstone 1857, p. 8).

Large-scale flood events are known to have occurred in all of the major southern Kalahari systems during the historical period. The Kuruman, for example, experienced extensive flooding in 1817–18, 1819–20, 1891–92, 1894, 1896, 1915, 1917–18, 1920, 1933–34, 1974–77 and 1988–89 (Fig. 15.2), whilst floods occurred in the Molopo in 1871, 1891–92, 1896, 1915, 1917, 1933–34, 1988 and, most recently, 1999–2000 (Nash 1996; Nash and Endfield 2002b). Records for the Auob and Nossob are more scanty, but the former is known to have flooded in 1933–34 and the latter in 1806, 1933–34, 1963 and 1987 (Nash 1996). The majority of historical flows appear to have had only limited effects upon contemporary valley morphology. There are, however, some notable exceptions. Floodwaters flowing from the Kuruman in 1894, for example, caused the Molopo to break its banks in the section of its valley below the Kuruman–Molopo confluence. Floodwaters were unable to follow the Molopo, which was partially blocked by sediment, and formed a large lake at Abiquasputs

Fig. 15.2 The Kuruman River (**a**) in flood near Donkerdraai Farm in 1988, which caused considerable disruption to infrastructure including (**b**) the road causeway at Grootdrink Farm (*Photographs* D. Nash). See Fig. 15.1 for locations of images

(Range 1912). This lake was refilled in 1934 as a result of widespread flooding which affected all four southern Kalahari drainage systems and is reported to have had a surface area of over 12,000 ha (Barrow 1974). On this occasion, floodwater extended as far as Noenieput, some 27 km south of Abiquasputs, whilst in 1976 flooding reached Springbok Vlei, 15 km north of Abiquasputs.

Sizeable flood events must also have occurred within southern Kalahari systems in prehistory. Wider sections of the Nossob within the Kgalagadi Transfrontier Park, for example, contain numerous abandoned channels and meander systems (Fig. 15.3a, b) which appear to have been cut off by floodwaters (Nash et al. 1994; Bullard and Nash 1998), whilst several episodes of Holocene flooding have been identified in the Kuruman (Shaw et al. 1992).

15.4 Dunes and Dunefields

Approximately 85 % of the south-west Kalahari between latitudes 23°S and 28° 20′S and longitudes 18°E and 22° 30′E is classified as dunefield (Fig. 15.1), with the remainder consisting of gently undulating sand sheets, valleys or pans (Thomas and Shaw 1991). The southern margin of the dunefield is marked by the Orange River (although some linear dunes occur south of the river around Upington), with dunes becoming less distinct north and east of the Nossob and Molopo valleys. Despite this being the driest part of the Kalahari (the area receives on average only 200–300 mm of rainfall per year), much of the dunefield is currently vegetated. This has a significant influence upon aeolian processes—most

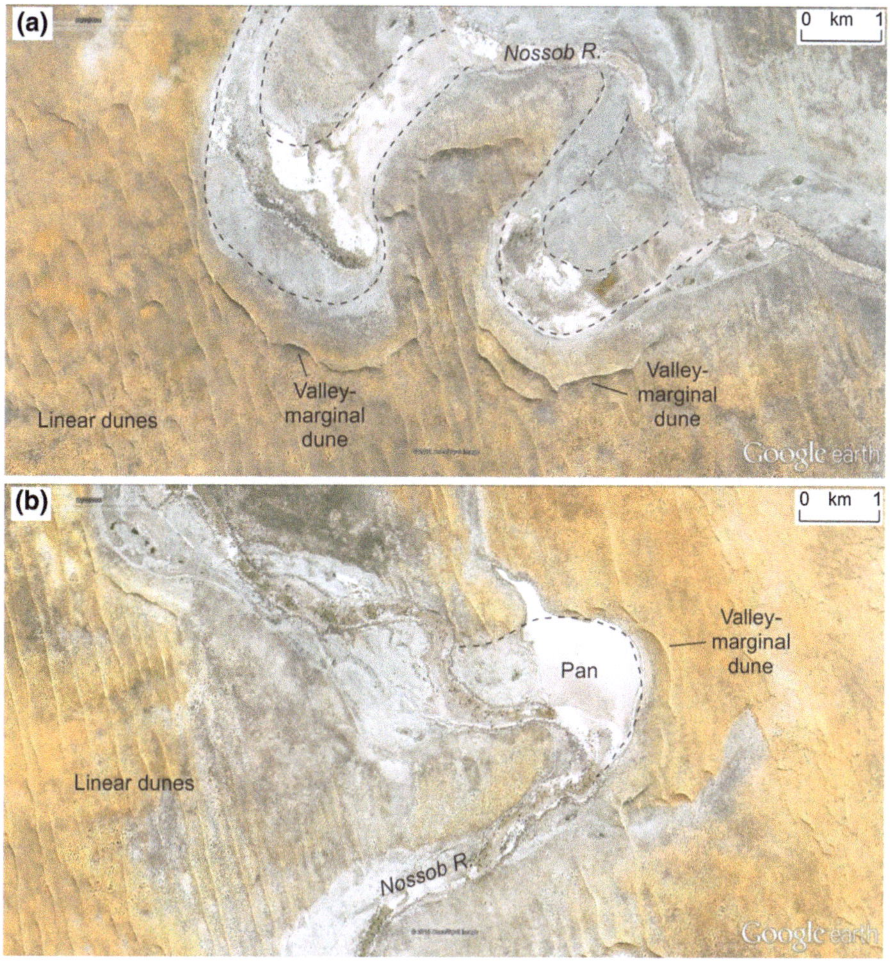

Fig. 15.3 Linear dunes, valley-marginal dunes, abandoned meanders (*dashed line*) and valley floor pans along the course of the Nossob River in the Northern Cape. See Fig. 15.1 for locations of images (**a**) and (**b**). Images courtesy of Google Earth (date of images—22 November 2006)

contemporary sand movement occurs only on dune crests, or in areas where vegetation has been cleared (for example as a result of fire, overgrazing or prolonged drought conditions), or during periods when winds are particularly strong.

Some 99 % of the south-west Kalahari dunefield is covered by orange–red coloured parallel to sub-parallel linear dunes (Fig. 15.3a). These have rounded (or in some areas sharp) crests that are orientated broadly north-west to south-east. Dune crests often support sparse clumps of grass which act as points of sediment accumulation, giving the dunes a hummocky appearance (Fig. 15.4). The linear dunes have been shaped over millennia by winds which have moved fine- to medium-grained sands in a broadly southeasterly direction towards the Orange River. In South Africa, linear dunes extend from the Kgalagadi Transfrontier Park as far south as Upington (Fig. 15.1). Within the Park, individual dunes range in height from 2 m to over 30 m above interdune areas, with some of the highest dunes occurring in the immediate vicinity of the Auob and Nossob valleys. Some of the largest dunes are up to 250 m wide, with interdune spacings of up to 2.5 km (Thomas and Shaw 1991), creating the impression of a gently undulating landscape.

The majority of linear dunes in the south-west Kalahari have a relatively simple form. However, spatial variations have been reported in a number of studies. Bullard et al. (1995), for example, identified five different classes of dune pattern within the dunefield (Fig. 15.5): discontinuous simple linear dune forms (Class 1); simple continuous forms that extend for several kilometres (2); compound forms where branching Y-junctions are common (3); compound forms with more obtuse angles between branches (4); and discontinuous dunes with no preferred orientation (5). Bullard and Nash (1998) also noted changes in dune pattern across drainage courses due to the influence of valleys upon sediment supply and local wind regime.

The remaining 1 % of dunes in the south-west Kalahari include a variety of forms. Most notable are small areas of parabolic dunes between the Auob and Nossob valleys, presumably developed as a result of wind activity following the disturbance of the vegetated or partially vegetated sand surface

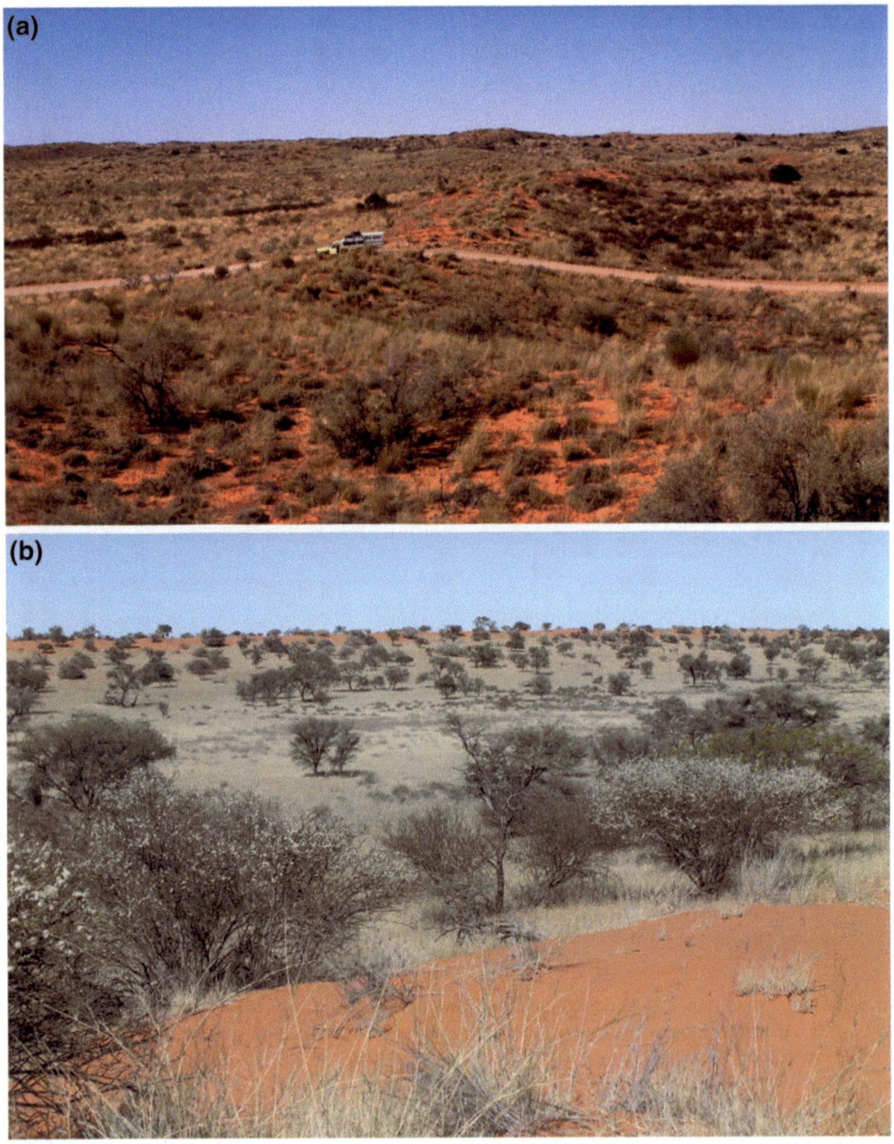

Fig. 15.4 Hummocky linear dune crest (*Photograph* D. Nash) (**a**) and interdune area within the south-west Kalahari dunefield (*Photograph* M. Baddock) (**b**). See Fig. 15.1 for locations of images

(Thomas and Shaw 1991). Isolated barchan dunes, climbing and falling dunes and flank dunes have also been documented. Crescentic (or 'lunette') dunes occur on many pan margins (see Sect. 15.5) and, from aerial photography, appear to act or have acted as sediment 'feeders' for linear dunes which continue downwind from them. Close to the Auob and Nossob, a variety of dune so far only identified in the south-west Kalahari can be seen. This landform, termed a valley-marginal dune (Fig. 15.3a, b), is found typically at the top of valley flanks, and has a narrow, elongate form with arcuate and straight planform elements which mirror the course of contemporary and ancient channels (Bullard and Nash 2000).

Luminescence dating of wind-blown deposits, including analyses of sediment incorporated within linear dunes, lunette dunes and sand sheets, has revealed a long history of aeolian activity across the Kalahari spanning at least the last 200 ka (thousand years) (Thomas and Burrough 2013). In the south-west Kalahari, phases of aeolian activity occurred around 104, 77 and 58–52 ka, with wind-blown sediments accumulating semi-continuously (as far as can be determined within the resolution of the luminescence dating technique) from around 35 ka to the present day.

15.5 Pan Environments

Pans are small, closed basins or depressions that contain seasonally flooded lakes and are characteristic globally of arid to semi-arid regions of low relief (see also Chap. 16). They are also of considerable importance as point water

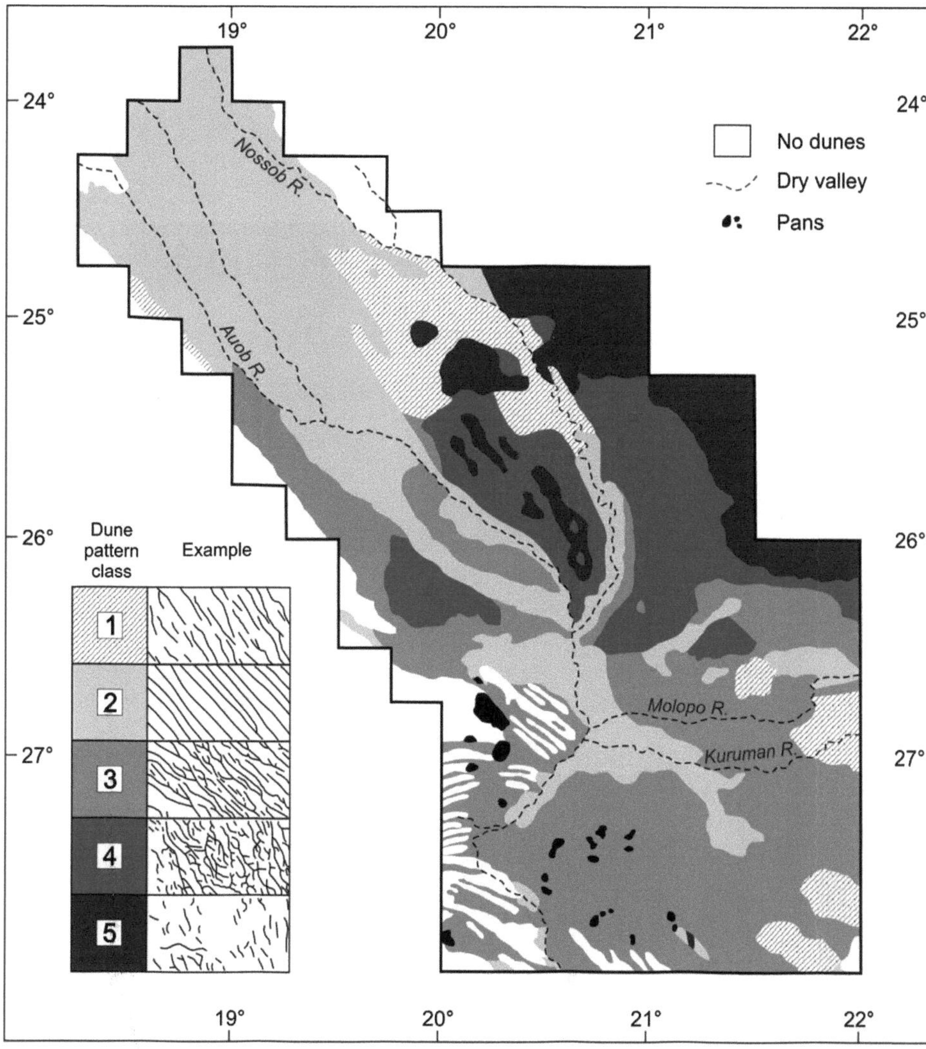

Fig. 15.5 Distribution of linear dune pattern classes in the south-west Kalahari dunefield (after Bullard and Nash 1998). See text for a description of the five dune patterns

sources for both humans and wildlife. Pans occur throughout the Kalahari, as far north as Zambia, but are a particularly significant component of the landscape in regions with (i) an annual rainfall of less than 500 mm and (ii) underlying rock or sediment that is susceptible to erosion by the wind (Goudie and Wells 1995). The southern Kalahari meets both of these criteria.

Two areas containing concentrations of pans occur within the southern Kalahari as defined here: (i) a broad zone extending from Upington, across the lower Molopo and northwards into Namibia (i.e. in the extreme south of Fig. 15.5) and (ii) a smaller area between Vryburg in North West Province and Tsabong in Botswana. Within these zones, pans occur as isolated landscape features (Fig. 15.6a) but also within depressions in the floors of dry valleys (Fig. 15.3b) and palaeodrainage lines, and in the corridors between linear dunes. The majority lack any surface inflow, although some may have short, poorly developed channels that supply run-off during major rainfall events. Pan sizes range from 1 to 16 km^2, with some reaching depths of 20 m below the surrounding ground surface (Thomas and Shaw 1991). Two of the largest isolated pans in the South African Kalahari—Hakskeen Pan and Koppieskraal Pan—can be accessed via good gravel roads immediately south-west of the Kgalagadi Transfrontier Park. Further examples within the Park include Polentswa Pan in the bed of the Nossob, and Bitterpan which is situated in the heart of the dunefield, midway between the Auob and the Nossob (4 × 4 access only).

Pans in the south-west Kalahari, particularly those around the Nossob–Molopo confluence, have a characteristic sub-circular, sub-elliptical or kidney shape and may have a preferred orientation parallel to the prevailing wind. Many are also associated with the presence of calcrete,

Fig. 15.6 The pan floor (**a**) and fringing lunette dune (**b**) at Witpan, a typical pan within the south-west Kalahari dunefield (*Photographs* D. Nash). See Fig. 15.1 for locations of images

which may form a rim around the pan margin and extend for several hundred metres beyond the pan depression. Large numbers of pans in the south-west Kalahari have one or more lunette dunes present on the leeward (typically southern) side of the pan margin in relation to the prevailing wind direction (Fig. 15.6a, b). Lunettes are commonly much paler in colour than the orange–red sands which blanket much of the surface of the south-west Kalahari. The largest lunettes can reach heights of 30 m relative to the pan floor. Some pans exhibit two lunette dunes, each with different sediment characteristics—an older outer dune composed predominantly of sand and a younger inner feature with a higher clay content. The differences between the two lunettes are thought to reflect climatic variability during their formation. The outer dune is likely to have formed as a result of phases of aeolian erosion of weathered sandy material from the pan floor during drier periods when the groundwater table was deeper. In contrast, the inner dune is a product of the deflation of finer sediments that accumulated during humid phases as a result of shallow water depositional processes (Thomas and Shaw 1991). To date, only a limited number of studies have attempted to investigate the environmental history of pan sediments and lunettes, but those that have been undertaken suggest considerable potential (Lawson and Thomas 2002; Telfer et al. 2009).

15.6 Summary

The combination of vegetated orange–red dunes, dry valleys and seasonal pans in the southern Kalahari creates a landscape with outstanding scientific and aesthetic value. From a geomorphological perspective, the region offers an almost unique environment in which to observe not only the operation of arid zone processes on individual landforms but also the integration of processes across a landscape. For example, topographic depressions such as pans and valleys provide sediment sources for dunes, but at the same time influence dune pattern; thick sequences of duricrusts influence the morphology of drainage systems, but at the same time are modified chemically as water flows through these systems. The protection of at least a small part of the Kalahari within the Kgalagadi Transfrontier Park means that this special landscape is not only managed for wildlife conservation but also for future geomorphological research and geotourism purposes.

References

Barrow B (1974) Song of a dry river. Purnell, Cape Town

Bullard JE, Nash DJ (1998) Linear dune pattern variability in the vicinity of dry valleys in the southwest Kalahari. Geomorphology 23:35–54

Bullard JE, Nash DJ (2000) Valley-marginal sand dunes in the south-west Kalahari: their nature, classification and possible origins. J Arid Environ 45:369–383

Bullard JE, Thomas DSG, Livingstone I, Wiggs GFS (1995) Analysis of linear sand dune morphological variability, southwestern Kalahari Desert. Geomorphology 11:189–203

Goudie AS, Wells GL (1995) The nature, distribution and formation of pans in arid zones. Earth Sci Rev 38:1–69

Grove AT (1969) Landforms and climatic change in the Kalahari and Ngamiland. Geogr J 135:191–212

Haddon IG, McCarthy TS (2005) The Mesozoic-Cenozoic interior sag basins of Central Africa: the Late-Cretaceous-Cenozoic Kalahari and Okavango basins. J Afr Earth Sc 43:316–333

Lawson MP, Thomas DSG (2002) Late Quaternary lunette dune sedimentation in the southwestern Kalahari Desert, South Africa: luminescence based chronologies of aeolian activity. Quatern Sci Rev 21:825–836

Leistner OA (1967) The plant ecology of the southern Kalahari. Botanical Research Institute Botanical Memoir 38. Republic of South Africa Department of Agricultural Technical Services, Pretoria

Livingstone D (1857) Missionary travels and researches in South Africa. John Murray, London

Moore AE, Larkin PA (2001) Drainage evolution in south-central Africa since the breakup of Gondwanaland. S Afr J Geol 104:47–68

Nash DJ (1996) On the dry valleys of the Kalahari: documentary evidence of environmental change in central southern Africa. Geogr J 162:154–168

Nash DJ, Endfield GH (2002a) A 19th century climate chronology for the Kalahari region of central southern Africa derived from missionary correspondence. Int J Climatol 22:821–841

Nash DJ, Endfield GH (2002b) Historical flows in the dry valleys of the Kalahari identified from missionary correspondence. S Afr J Sci 98:244–248

Nash DJ, McLaren SJ (2003) Kalahari valley calcretes: their nature, origins, and environmental significance. Quatern Int 111:3–22

Nash DJ, Thomas DSG, Shaw PA (1994) Timescales, environmental change and dryland valley development. In: Millington AC, Pye K (eds) Environmental Change in Drylands: Biogeographical and Geomorphological Perspectives. Wiley, New York, pp 25–42

Range P (1912) Topography and geology of the German South Kalahari. Trans Geol Soc South Africa 15:63–73

Shaw PA, Thomas DSG, Nash DJ (1992) Late Quaternary fluvial activity in the dry valleys (mekgacha) of the middle and southern Kalahari, southern Africa. J Quat Sci 7:273–281

Telfer MW, Thomas DSG, Parker AG, Walkington H, Finch AA (2009) Optically stimulated luminescence (OSL) dating and palaeoenvironmental studies of pan (playa) sediment from Witpan, South Africa. Palaeogeogr Palaeoclimatol Palaeoecol 273: 50–60

Thomas DSG, Burrough SL (2013) Luminescence-based dune chronologies in southern Africa: analysis and interpretation of dune database records across the subcontinent. Quatern Int. doi:10.1016/j.quaint.2013.09.008

Thomas DSG, Shaw PA (1991) The Kalahari environment. Cambridge University Press, Cambridge

Thomas DSG, Shaw PA (2002) Late Quaternary environmental change in central southern Africa: new data, synthesis, issues and prospects. Quatern Sci Rev 21:783–797

The Western Free State Panfield: A Landscape of Myriad Pans and Lunettes

16

Peter Holmes

Abstract

The descriptor *Western Free State Panfield* covers the greatest concentration of pans in southern Africa. Many Free State pans boast fringing lunette dunes on their southern and south-eastern margins. Lunettes derive their name from their sickle-moon shape, yet the environmental requirements which promote and sustain the formation of pan-lunette sequences are still not fully understood. In the western Free State, pans form almost exclusively on the shale substrate of the Ecca Group, suggesting a significant measure of geological control. An apparent pattern of pan orientation exists in agreement with palaeodrainage lines. This supports the notion of bedrock influence on drainage, with drainage in turn playing a direct role in pan formation. Lithological (rock type) control on the formation of pans in the western Free State is, clearly, important. This chapter describes the morphology of pans and lunettes (the two features should, arguably, not be separated) and reviews the origins of their formation. Finally, pans as a resource are examined, and their significance as a key element of the Free State landscape and a proxy for environmental change is assessed.

Keywords

Aeolian • Lunette • OSL dating • Pan • Sediment • Free State

16.1 Introduction

Pans (Spanish: *playas*) are enclosed depressions found in many dryland areas of the world and are widespread in parts of southern Africa. Shaw (1988) identified eight areas of pan concentration, six of which occur in or immediately adjacent to South Africa (Fig. 16.1). Pans generally consist of endorheic, flat, periodically inundated, unvegetated basins, typically up to several km^2, or even larger, in area. A range of geomorphic processes may contribute to their development and, in some regions of southern Africa, they occur along now-disrupted palaeodrainage lines (Goudie and Thomas 1986; Goudie and Wells 1995). Lunette dunes, which display a characteristic crescentic morphology, frequently flank the downwind margins of pans. The lateral extent of lunettes is typically a function of the individual pan's downwind shoreline length. Lunettes, or suites of lunettes extending for a kilometer or more, are not uncommon. Typically, lunette crests reach between 2 and 4 m in height above the pan shoreline. Lunettes comprise sediment commonly regarded as having been derived from the basin floor, though material may be partially sourced from upwind of the depression (Telfer and Thomas 2006). Lunette sediments comprise material ranging in size from clay to sand (Lawson and Thomas 2002); the former attributed to the direct pan-floor deflation of clay pellets during times of water table fluctuation, and the latter to deflation of littoral sand moved to the pan margin by wave action (Bowler 1973). Suites of lunettes occur at some pans, with the oldest situated farthest from the contemporary pan. Lunette dunes are important palaeoenvironmental archives in regions where organic preservation is low, and thus many previous studies have used lunettes to reconstruct changes in aridity and wind direction (see Thomas and Shaw 2012).

P. Holmes (✉)
University of the Free State, Bloemfontein, South Africa
e-mail: holmespj@ufs.ac.za

S. Grab and J. Knight (eds.), *Landscapes and Landforms of South Africa*,
World Geomorphological Landscapes, DOI 10.1007/978-3-319-03560-4_16,

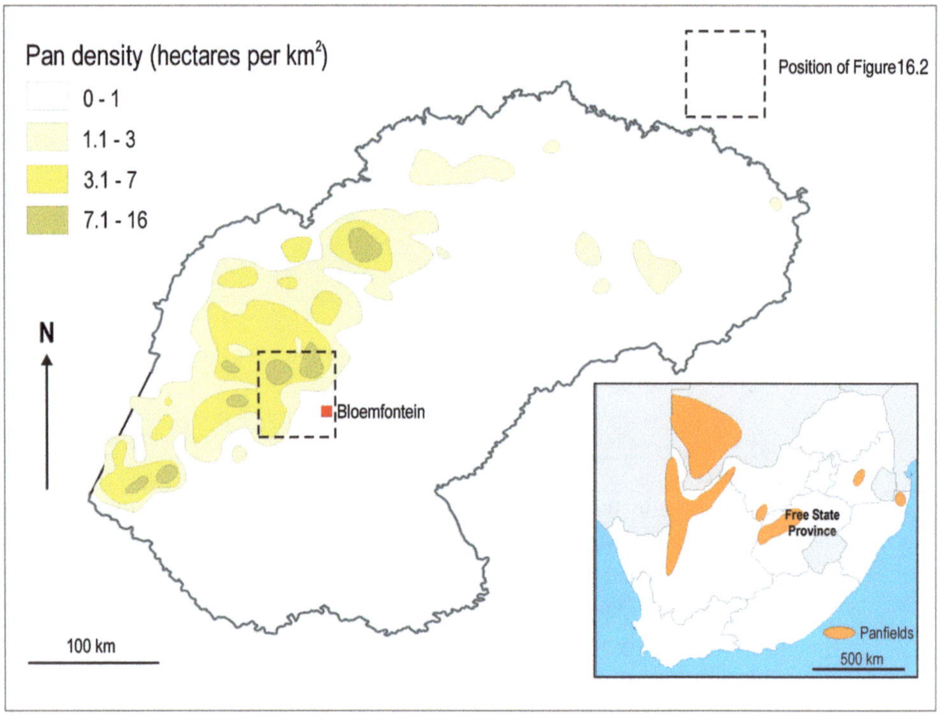

Fig. 16.1 Panfields of southern Africa, and the Western Free State Panfield (adapted after Holmes et al. 2008)

The high plains (mean altitude ~1,400 m a.s.l.) of the western Free State Province (hereafter referred to as the Free State) exhibit the greatest concentration of pans in South Africa. These pans typically cover between one and seven hectares per km^2 of land area, with accumulative areas of coverage reaching up to 16 hectares per km^2 in places (Fig. 16.1). At many of these pans, lunette dunes are common on southern and south-eastern shorelines. The Free State is situated within South Africa's summer rainfall area, and its western half currently receives less than 550 mm of rain per annum. It is a transition from the dry Highveld Grassland to Kalahari Hardveld Bushveld vegetation types. As such, the Free State is potentially climatically sensitive to environmental changes which might be recorded in pan-lunette suites. It should also be noted that, although the present-day Kalahari is situated to the northwest, during periods of high aridity, the western Free State would have been more proximal in terms of the environmental conditions it experienced (Thomas and Shaw 2002).

16.2 Climate and Geology

The Free State falls exclusively within the catchments of the westward-draining Vaal and Orange Rivers which form the northern and southern borders of the province. The Drakensberg defines the eastern margin of the Free State. Maximum rainfall associated with summer Indian Ocean monsoon, occurs in February and March. Contemporary winds are highly variable, tending mainly north to north-east with north-westerly winds associated with the passage of cold fronts, typically in late winter, with October being the windiest month (Wiggs and Holmes 2011).

The western Free State is dominated by late Paleozoic and Mesozoic sedimentary rocks of the Karoo Supergroup. Lithological control of the landscape is apparent; sedimentary rocks (mostly shales) of the Ecca Group dominate the western Free State. Weathering and erosion of these rocks result in the typical flat-to-undulating landscape of the west-central Free State, with occasional flat-topped, dolerite-capped mesas and buttes. Quaternary deposits are widely distributed, comprising calcretes and unconsolidated sandy alluvial and colluvial deposits. Surficial veneers of sand on parts of the western Free State landscape are likely of aeolian (wind-derived) origin (Holmes and Barker 2006).

16.3 Pans and Lunettes

Pans have enjoyed considerable attention in the South African literature for more than half a century. Citing earlier work where, *inter alia*, drainage derangement, wind erosion and removal of dissolved salts and clay by animals have been suggested as causes of pan formation, Le Roux (1978) observed that the drier western to north-western Free State displays the highest concentration of pans (Figs. 16.1 and 16.2). This zone largely coincides with the Ecca Group geology, suggesting that the shales of the Ecca Group are

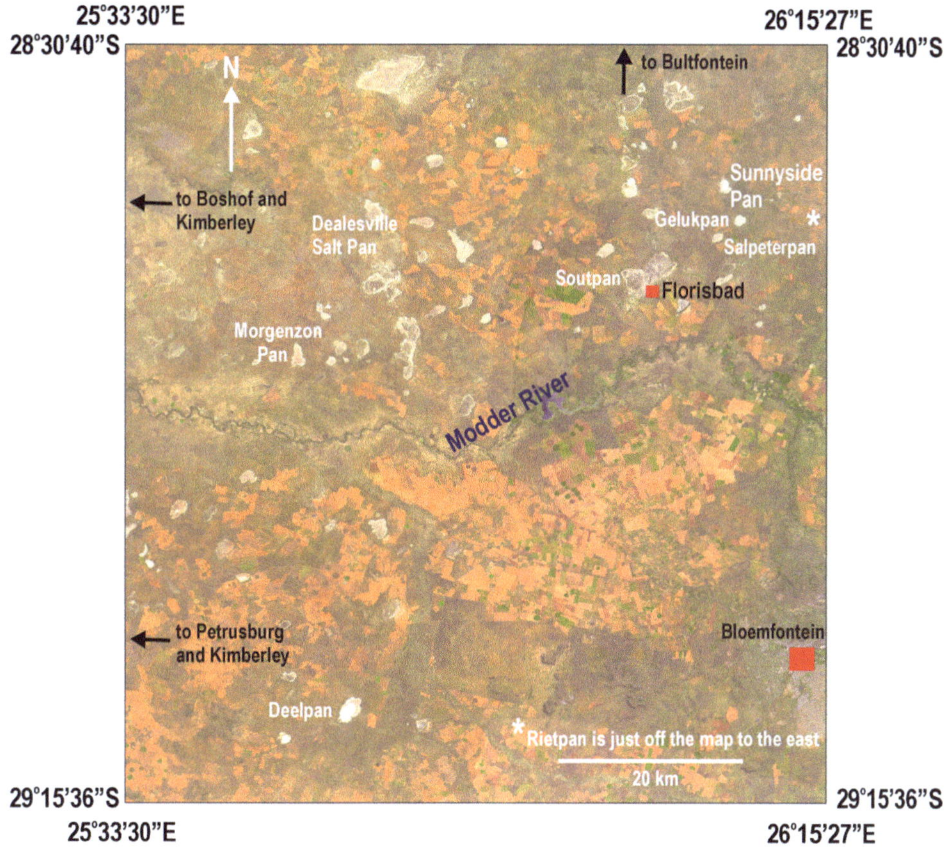

Fig. 16.2 Satellite image of the section of the Western Free State Panfield in the immediate vicinity of Bloemfontein (see also Fig. 16.1)

particularly conducive to pan formation. A schematic representation of the variables which play a role in pan formation is presented in Fig. 16.3 (after Goudie and Thomas 1985).

The presence and formation of lunette dunes associated with pans in the Free State was broadly discussed by Goudie and Thomas (1986). Additionally, Scott and Brink (1992) investigated the palynology and palaeontology of lunette dunes at Deelpan and Florisbad (Fig. 16.2). Compared to lunettes from other regions of South Africa and elsewhere, especially the extensive Kalahari region, where chronologies exist of lunette accumulation based on optically stimulated luminescence (OSL) dating, comparatively little was known about the temporal aspects of pan deflation and sediment accumulation in pan-fringing lunettes within the western Free State. This has since been addressed at both sub-regional and site specific levels (e.g. Holmes et al. 2008; Rabumbulu and Holmes 2012).

16.4 Pan-Lunette Sites

Driving through the western Free State, it is inevitable that the visitor will encounter one or more pans. In most instances, fringing lunettes will be readily apparent (Fig. 16.4). The most detailed studies (distribution, morphology) of pan-lunette sequences in the western Free State include: Morgenzon Pan (28°49′36″S; 25°42′34″E); Sunnyside Pan (28°39′15″S; 26°08′53″E); Salpeterpan (28°42′16″S; 26°08′07″E); Deelpan (29°11′11″S; 25°45′27″E) and Rietpan (28°40′30″S; 26°18′25″E) (Holmes et al. 2008). Their positions are indicated on the accompanying satellite image (Fig. 16.2).

16.5 Pan and Lunette Characteristics

Typically, pans appear as large (up to 1 km^2, and in some cases larger), flat, oval-shaped, desolate amphitheatres, devoid of vegetation. They may be completely inundated with water during late summer (Fig. 16.4). Most often, however, the pan surface will comprise a dry, wind-swept crust of (frequently saline) mud (Fig. 16.5a). At Soutpan and the Dealesville Salt Pan, extensive salt mining operations criss-cross and pockmark the pan surfaces (Fig. 16.5b). Other pans, such as Morgenzon and Deelpan, display pristine surfaces. Pan surfaces are harsh, inhospitable environments with high windspeeds and extreme winter minimum temperatures (−10 °C). The former are due to the exposed

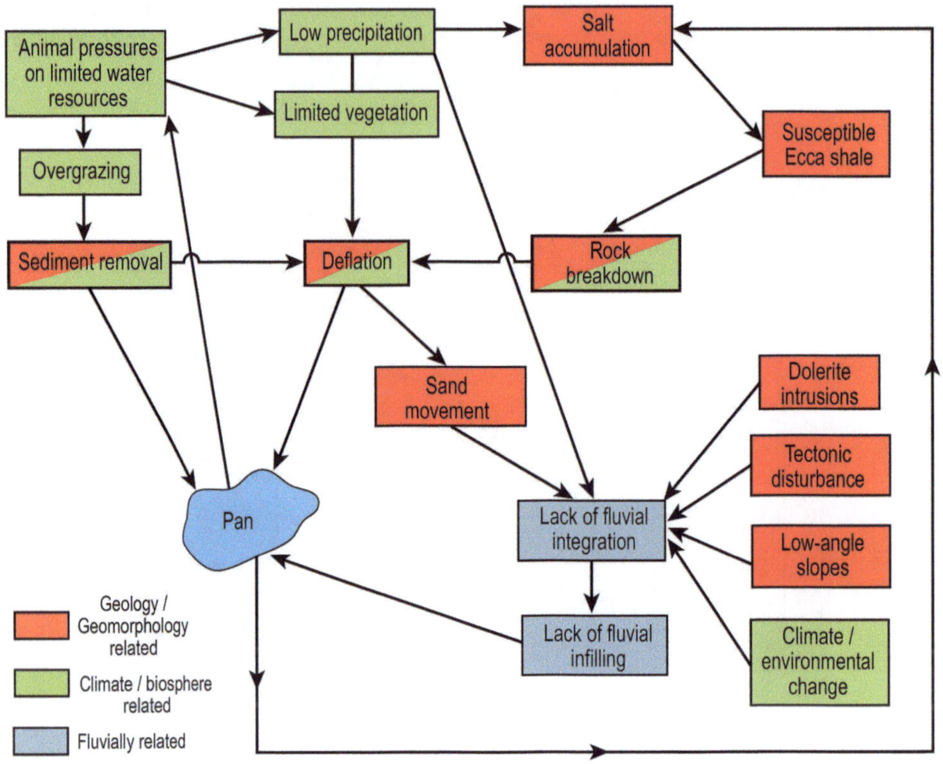

Fig. 16.3 The formation of pans (adapted after Goudie and Thomas 1985)

Fig. 16.4 Panoramic view of Deelpan. Lunette dunes (*inset*) are visible in the background (*Photograph* P. Holmes)

surface, with a significant fetch (the unobstructed distance over the pan surface over which the wind blows) and the latter because of the typical pan position, located in a depression or hollow occupying a low point in the local topography. Summer temperatures on pan surfaces habitually exceed 40 °C. Judging from the local topography, pan floor sediment accumulations appear to be of limited depth (metres, rather than tens of metres). No apparent consistency between pan long-axis length and wind direction is evident. However, there may be a relationship between long axis length and palaeodrainage, but this still requires substantiation. Various hypotheses, including animal pressure on limited water resources, overgrazing, selective wind deflation, limited vegetation, geological substrates and soil accumulation, have been suggested for pan formation (Thomas and Shaw, 2012). In virtually all cases, the lunettes on the downwind margins of pans have been dissected by fluvial erosion, as surface runoff returns to the pan. Most lunettes are at least partially vegetated and may (or may not) display some evidence of aeolian structure, such as cross-bedding (Fig. 16.5c). There are often colour differences within the sedimentary facies (layers, or strata) exposed within the lunettes. There is a noticeable regularity of the occurrence of lunettes on the southern to south-eastern pan

Fig. 16.5 **a** Pan surface, showing desiccation and cracking of clay; **b** salt works on Soutpan; **c** gullied and dissected lunette at Sunnyside Pan, note evidence of sedimentary structure [bedding], colour changes, and vegetation on the lunette. Holes are sample points where sediment has been systematically extracted down the profile for OSL dating; **d** fence-line dunes near Florisbad. Remnants of the original barbed wire fence are to be found in this dune, indicating that the dune post-dates ~1890 when barbed wire was introduced to the Free State (*Photographs* P. Holmes)

margins, consistent with a dominant current and palaeo-north-westerly wind direction over central southern Africa.

At Morgenzon Pan and at Deelpan (Fig. 16.2), there are secondary, less prominent (or remnant) lunettes farther from the respective pans than their flanking lunettes. There is no apparent evidence of multiple lunettes at the other sites mentioned, but this certainly does not preclude suites of lunettes at other western Free State pan sites.

While western Free State pan surfaces typically comprise sediment in the silt to clay size fraction, their accompanying lunettes comprise predominantly well-sorted sand, with fine sand dominating the sand fraction. These lunettes are alkaline environments, with pH values ranging from a low of 7.32 to a high of 10.22, while the presence of calcium carbonate is ubiquitous. From a palaeoenvironmental perspective, lunette dunes are, arguably, of greater importance than pans. The western Free State lunettes preserve a record of discontinuous sedimentation through the past 18 ka, as revealed by OSL dating. In the broadest sense, this is comparable with the record of lunette accumulation from the southern Kalahari, where lunette dunes are a widespread localised landscape feature. Lunette development demonstrates that there has been sufficient wind energy to construct dunes when sediment has been available from local sources. Accumulation rates have been significant; 2 m of accumulation has occurred at the Salpeterpan lunette in the last 280 years, 1.4 m at Deelpan in the last 100 years, 1.3 m at Sunnyside Pan in the last 200 years, and up to 1.3 m at the Morgenzon Pan lunette in the last 330 years. The youngest phase of accumulation is accounted for in terms of possible human impacts, including agriculture in the western Free State, since colonial times (Holmes et al. 2012).

Gullying of lunettes by local drainage returns sediment from the lunette to the pan floor, from where it is once again available for entrainment by wind. This process of sediment cycling has an impact on the preservation of sediments in the lunette from earlier phases of accumulation. Lunette dune development is not necessarily an indication of aridity: it indicates sediment availability to the wind and conditions that allow sediment moved from the pan to accumulate on its downwind margin.

Lunette dune accumulation at the western Free State pan sites at 10–12 ka corresponds with a period of very dry conditions recorded at Kathu Pan in the southern Kalahari (Holmes et al. 2008). More recent lunette development 1–2 ka ago corresponds with dry conditions recorded in pollen records (Scott 1988). The marked phase of lunette construction from 3.5 to 5.0 ka presents an interesting challenge given the wet conditions inferred from spring deposits and peat accumulations on the floor of Deelpan at 3.5 and 4.5 ka cal BP. However, these are recorded as short events and coincide with a generally drier period recorded at Kathu Pan. Overall, when all the evidence is taken together, there is a marked impression that lunette accumulation has occurred during dry periods in the southern African interior.

16.6 Florisbad

Florisbad (Fig. 16.2) is a unique geomorphic site, in that it combines elements of pan-lunette topography with a natural spring and a rich archaeozoological heritage (Brink 1988), and thus deserves special mention. Florisbad comprises a spring mound, situated to the south-east of Soutpan (Fig. 16.2), which forms a local base-level. The crest of the mound is 32 m above the floor of Soutpan, while the diameter of the mound is some 500 m. Recent research (e.g. Douglas et al. 2010; Rabumbulu and Holmes 2012) has confirmed the close association between Soutpan and Florisbad. A number of hypotheses have been put forward for the formation of the Florisbad spring mound and the excellent state of preservation of faunal remains at the site; these are summarised by Douglas et al. (2010). Because of its geomorphological complexity, including the presence of an active spring, the origins of the spring mound continue to be debated. A recent suggestion is that the mound comprises aeolian sand, and that it represents a well preserved lunette dune in close proximity to the natural spring, which abetted the formation and stabilisation of the lunette (Douglas et al. 2010).

16.7 The Significance of Western Free State Pans and Lunettes

Sediment mobility and lunette sedimentation during the last few centuries continue to pose questions. In the Free State, enhanced sediment mobility may coincide with the period of colonisation by Europeans, and the introduction of farming, particularly when considering the current potential for aeolian erosion on ploughed land in the region (Wiggs and Holmes 2011). Could such human-induced causes have resulted in a period of landscape instability, and enhanced aeolian activity during the historical era? This question remains only partially answered (Holmes et al. 2012), yet it is worth noting that this phase of dune building is also recorded in the Kalahari. Fence line dunes (Fig. 16.5d) in the western Free State are currently a feature of wind erosion under tillage (Holmes et al. 2012). These dunes form when wind-blown soil from an adjacent field accumulates against fence posts. Ultimately, a linear dune forms along the length of the fence, in some cases burying the original fence. Whether these modern dunes are an analogue for landscape instability and enhanced aeolian activity during the colonial period is debatable.

The Western Free State Panfield remains an intriguing part of the rich geomorphic diversity of South Africa. The pans, despite their brackish water, provided an important source of drinking water for wildlife during former times. They currently support populations of various water birds. Currently, pans such as Soutpan and the Dealesville salt pan support local livelihoods through the exploitation of their salt reserves. Above all, the potential ability of pans and their fringing lunettes to: (a) contribute to a better understanding of environmental/climatic change on the South African highveld, and (b) provide a regional geo/eco-tourist attraction, confirms their considerable geomorphic value.

16.8 Summary

The Western Free State Panfield is an important geomorphological component of the arid to semi-arid western part of South Africa. Lunette dunes are common to many of the pans within this region. The occurrence of lunettes on the south-eastern margins of Free State pans is consistent with established records of palaeocirculation and wind direction over central southern Africa during the Late Quaternary. Phases of lunette accretion since the Last Glacial are also in good agreement with the findings from pan-fringing lunettes in the south-west Kalahari. Whilst it may be tempting to ascribe lunette construction primarily to aridity, a complex combination of factors, not least of which is sediment availability under a consistent wind regime, is necessary for lunette construction. Currently, aeolian conditions in the western Free State are such that sand accretion along fence lines is a marked feature of the landscape. However, lunette dunes appear to be in a phase of degradation and are susceptible to fluvial erosion and sediment recycling back into their respective pans. At the mesoscale, Free State lunettes appear to be an artefact of environmental conditions which post-date the Last Glacial. Detailed OSL dating sequences have provided good temporal resolution for selected western Free State lunettes. However, more recent microscale studies detailing aeolian process dynamics seem a necessary geomorphic contribution to better understand the origins of these pans and lunettes.

Acknowledgements I thank my colleagues David Thomas, Mark Bateman, Johan Loock, Matt Telfer, Charles Barker, Giles Wiggs, Martin Lawson, Mulalo Rabumbulu, Rodney Douglas and Marian Tredoux for fruitful research collaboration in the western Free State. Thanks to the referees who provided helpful suggestions which considerably improved this contribution. This chapter is dedicated to my student and friend, Rod Douglas who passed away in early 2014, and from whom I learnt so much.

References

Brink JS (1988) The taphonomy and palaeoecology of the Florisbad spring fauna. Palaeoecol Afr 19:169–179

Bowler JM (1973) Clay dunes their occurrence formation and environmental significance. Earth Sci Rev 9:315–338

Douglas RM, Holmes PJ, Tredoux M (2010) New perspectives on the fossilization of faunal remains and the formation of the Florisbad archaeozoological site, South Africa. Quatern Sci Rev 29:3275–3285

Goudie AS, Thomas DSG (1985) Pans in southern Africa with particular reference to South Africa and Zimbabwe. Zeitschrift für Geomorphologie NF29: 1–19

Goudie AS, Thomas DSG (1986) Lunette dunes in southern Africa. J Arid Environ 10:1–12

Goudie AS, Wells GL (1995) The nature, distribution and formation of pans in arid zones. Earth-Sci Rev 38:1–69

Holmes PJ, Barker CH (2006) Geological and geomorphological controls on the physical landscape of the Free State. S Afr Geogr J 88:3–10

Holmes PJ, Bateman MD, Thomas DSG, Telfer MW, Barker CH, Lawson MP (2008) A Holocene-late Pleistocene aeolian record from lunette dunes of the western Free State panfield, South Africa. Holocene 18:1–13

Holmes PJ, Thomas DSG, Bateman MD, Wiggs GFS, Rabumbulu M (2012) Evidence for land degradation and aeolian sediment in the west-central Free State Province, South Africa. Land Degrad Dev 23:601–610

Lawson MP, Thomas DSG (2002) Late Quaternary lunette dune sedimentation in the southwestern Kalahari desert, South Africa; luminescence based chronologies of aeolian activity. Quatern Sci Rev 21:825–836

Le Roux JS (1978) The origin and distribution of pans in the Orange Free State. South African Geographer 6:167–176

Rabumbulu M, Holmes PJ (2012) Depositional environments of the Florisbad Spring site and surrounds; a revised synthesis. S Afr Geogr J 94:191–207

Scott L (1988) Holocene environmental change at western Orange Free State pans, southern Africa, inferred from pollen analysis. Palaeoecol Afr 19:109–118

Scott L, Brink JS (1992) Quaternary palaeoenvironments of pans in central South Africa: palynological and palaeontological evidence. S Afr Geographer 19:22–34

Shaw PA (1988) Lakes and pans. In: Moon BP, Dardis GF (eds) The geomorphology of southern Africa. Southern Book Publishers, Johannesburg 120–140

Telfer MW, Thomas DSG (2006) Complex Holocene lunette dune development South Africa; implications for paleoclimate and models of pan development in arid regions. Geology 34:853–856

Thomas DSG, Shaw PA (2002) Late Quaternary environmental change in central southern Africa; new data synthesis issues and prospects. Quatern Sci Rev 21:783–797

Thomas DSG, Shaw PA (2012) Terminal basins: lacustrine and pan systems. In: Holmes PJ, Meadows ME (eds). Southern African geomorphology; recent trends and new directions. SUN, Bloemfontein, pp 169–187

Wiggs GFS, Holmes PJ (2011) Dynamic controls on wind erosion and dust generation on west-central Free State agricultural land, South Africa. Earth Surf Proc Land 36:827–838

The Sterkfontein Caves: Geomorphology and Hominin-Bearing Deposits

17

Dominic Stratford

Abstract

The Sterkfontein Caves, located in the south-west of the Cradle of Humankind in Gauteng is the world's richest *Australopithecus*-bearing locality. The fossil-bearing cave deposits represent a more recent instalment of a history spanning 2.6 Ga, from the deposition of the karst-hosting dolomites, to the commercial exploitation of the caves by lime miners in the early twentieth century. The location and morphology of the caves is a result of lithological variation within the two host dolomite formations, multiple and complex phases of karstification and infilling of the resultant solution cavities over the two billion years since the dolomite deposition, and consistently active local tensional joint and fault systems. Where vadose collapse has opened the caves to the landscape, a broad range of geomorphological processes has created dynamic sedimentary environments with complex stratigraphic histories.

Keywords

Karstic caves • Malmani dolomites • Cave deposits • *Australopithecus*

17.1 Introduction

The Cradle of Humankind World Heritage Site covers an area of approximately 800 km^2 in the north-west of Gauteng Province (Fig. 17.1). The boundaries of the Cradle of Humankind (hereafter Cradle) partially enclose the exposed and karstified dolomitic limestone, which hosts the famous hominin-bearing caves. The densest collection of excavated fossil-bearing sites, including Sterkfontein, is located in the south-western area of the Cradle near the Blaauwbank river valley (Fig. 17.1). The site occupies the top of a low hill about 300 m south of, and rising 50 m above (1,491 m a.s.l.), the Blaauwbank River. Palaeoanthropological attention was brought to Sterkfontein after lime mining activities had de-roofed the uppermost chamber, exposing the fossil-rich cave infill. As a by-product of the blasting of the calcite deposits, large fossil-bearing dumps were created. It was in one of these dumps that the first hominin fossil was found in 1936. The site remains one of the most prolific and long-studied fossil hominin localities in the world, yielding over 750 hominin specimens. Key finds include the first adult cranium of an *Australopithecus* found in 1936, the first complete adult *Australopithecus* cranium, affectionately known as Ms. Ples, and the most complete skeleton of an *Australopithecus* yet discovered (Clarke 2013). This chapter reflects on: (1) the hypothesized geomorphic evolution and formation of the Sterkfontein Caves, (2) the cave geomorphology and interned fossil-bearing sediments, and (3) the significance of the caves to palaeoanthropology and geotourism.

17.2 Geological Context

The Sterkfontein karst caves formed within the Neoarchean dolomites of the Malmani Subgroup which were deposited under epeiric (inland) marine conditions (Eriksson et al. 1995) between 2,550 and 2,423 Ma and reach a thickness of 1,450 m close to Sterkfontein (Truswell and Eriksson 1975). A range of tidal, deep-water and chemical depositional environments accumulated the dolomite and interstratifying

D. Stratford (✉)
University of the Witwatersrand, Johannesburg, South Africa
e-mail: dominic.stratford@wits.ac.za

S. Grab and J. Knight (eds.), *Landscapes and Landforms of South Africa*,
World Geomorphological Landscapes, DOI 10.1007/978-3-319-03560-4_17,

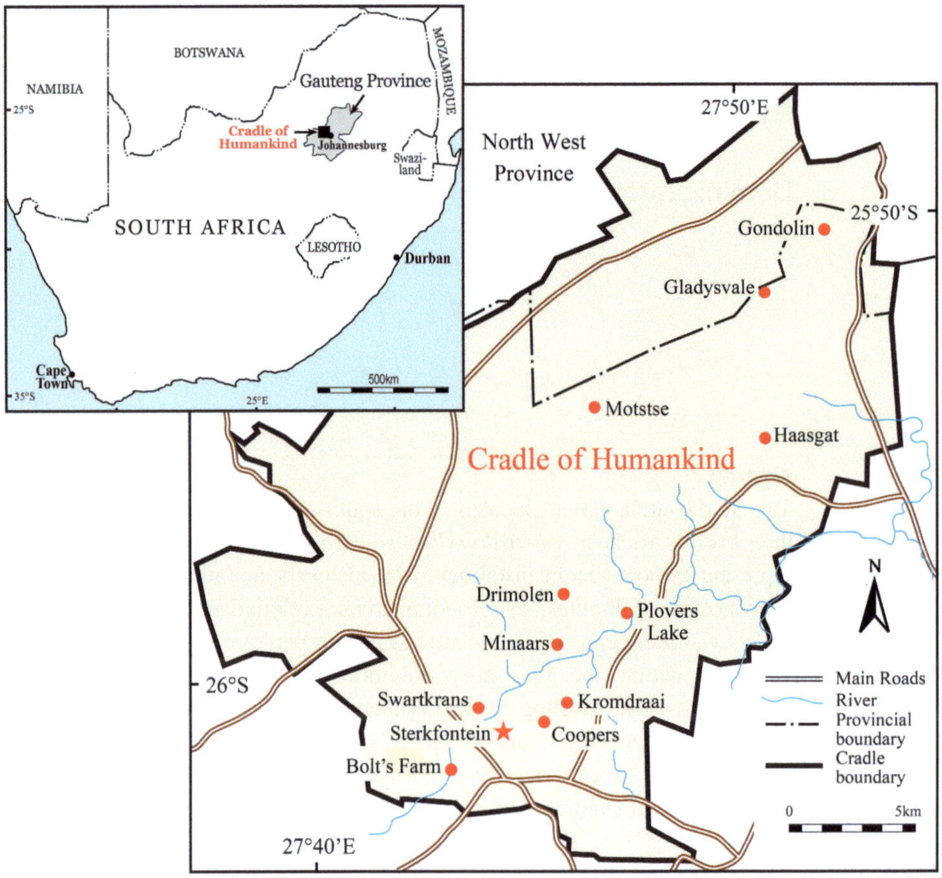

Fig. 17.1 Sterkfontein and other key fossil sites within the Cradle of Humankind, UNESCO World Heritage property, Gauteng, South Africa

chert beds within a general carbonate ramp model framework (Sumner and Grotzinger 2004). The Sterkfontein caves have formed within and across the boundaries of the two lower formations of the Malmani Subgroup (Fig. 17.2). The Oaktree Formation (basal member) is poor in chert beds and is overlain by the Monte Christo Formation, comparatively rich in chert beds (Martini et al. 2003). The difference in chert bed density between these two formations has strongly influenced the formation and geomorphology of the current cave system.

In the Sterkfontein area, early karst caves, formed during the early subaerial exposure of the dolomites, and open faults developed in the Monte Christo Formation were filled by chert breccias of the Rooihoogte Formation (Dirks and Berger 2013). This formation represents the proximal and medial portions of alluvial fans formed from transported weathered products of the uplifted dolomite landscape (Catuneanu and Eriksson 2002). Remnants of this chert breccia palaeokarst fill are found in close association with the cave sites of Sterkfontein, Swartkrans, Kromdraai, Coopers Cave and Gladyvale. Where this association is found, the unconformities formed by the chert breccia-filled palaeokarst provided the focus for the next generation of karstification, as the dolomites were again exposed and weathered.

Martini et al. (2003) suggest that the start of the Sterkfontein cave formation took place between 18 and 5 Ma, inferred from Miocene uplift of the area, erosion and exposure, and the age of oldest deposits yet found in the caves. Alternatively, Dirks and Berger (2013) propose that exposure of the dolomites took place after a much more recent erosion of overlying Karoo sediments ($\sim$2 Ma).

17.3 Karstification

The hyperphreatic model of cave formation has received great support over the history of geological research at Sterkfontein. The classic hyperphreatic model proposes that lowering of base level will result in the development of successively deeper parts of the cave network as dissolution is greatest within the upper 10 m of the local groundwater level (Martini et al. 2003). This model supports ideas of a lowering water table level related to uplift and erosion

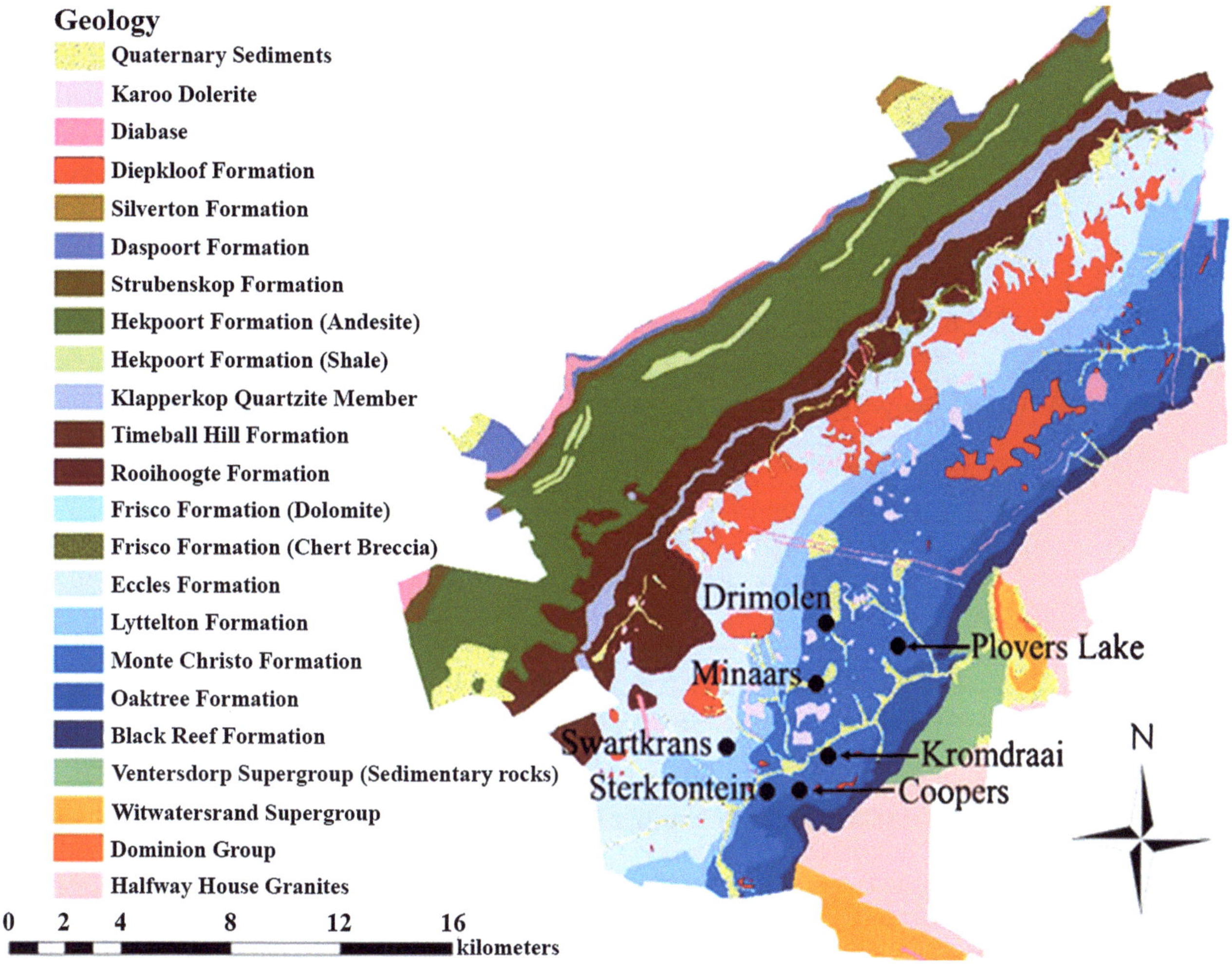

Fig. 17.2 Geological map of the Cradle of Humankind (adjusted from Obbes 2000)

cycles. However, Wilkinson (1973, 1983) proposed a 'deep phreatic' karstification process, suggesting that the entire vertical extent of the cave system had developed prior to the caves opening to the surface. This is based on geomorphological mapping of the cave network and tracking vadose autogenic and allogenic (internally and externally derived) colluvial deposits to the base of the cave system. Another alternative could be the hypogenic origin of the cave–i.e. its formation through rising deep water flow (likely thermal) (Martini et al. 2003; Klimchouk 2007). A more recent suggestion has been the in situ chemical alteration of the dolomite into 'ghost rock' (Bruxelles et al. 2009; Dubois et al. 2014). This process creates pseudophreatic karst morphologies, differing in the process of dissolution and final removal of the dolomite. Residual dolomite is removed as the groundwater level drops, leaving a vertically fully developed karst network as the caves open to the surface through vadose collapse. Low penetration of meteoric recharge and near-static groundwater also fits this formation model.

Identification of the specific development model for the Sterkfontein Caves has important implications for the age of the interred fossil and artefact-bearing deposits, a subject that remains a source of significant and continued debate (e.g. Herries and Shaw 2011; Stratford et al. 2014). The hyperphreatic model proposes that fossiliferous deposits found in the deepest areas must be younger than, or reworked from, older chambers above. Recent renewed geomorphological and stratigraphic work at Sterkfontein supports previous observations of early allogenic sediments being deposited at the base of the system (Wilkinson 1973), but has not yet succeeded in providing new absolute dates for these early sediments (Stratford et al. 2014).

17.4 Cave Geomorphology

The vertical depth of the karst network is difficult to estimate as passages descend below the groundwater table, but has been proposed to reach a depth of 79 m (Moen and Martini

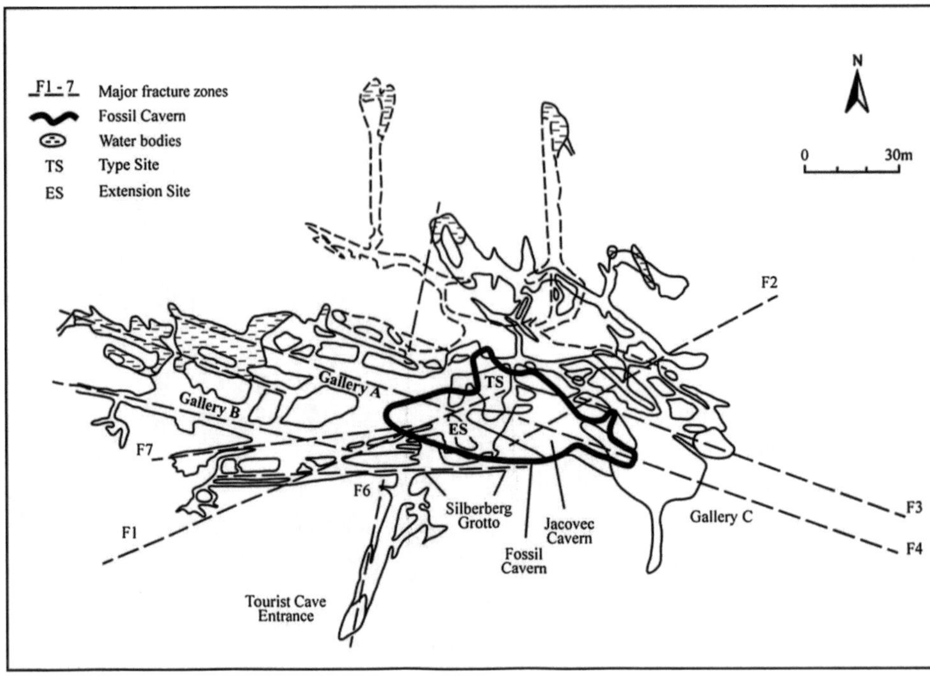

Fig. 17.3 Geomorphological plan of the Sterkfontein cave system. Major controlling joints and faults are shown in relation to the upper level (Fossil Cavern; *bold line*) and the lower level (subterranean area; *thin lines*)

1996). Lateral karst developmental controls are guided by dominant and subordinate fracture systems within the heavily fractured dolomites (Fig. 17.3). A dominant fracture system runs roughly east–west, and a subordinate system runs north–south. The influence of these fractures and joints can be seen in the perpendicular intersections of the major passages and galleries. The cave network is limited to an area 350 m east–west and 250 m north–south. In the phreatic system (under the water table), faults enable water penetration within the vertical fault planes, forming tall, narrow passages in chert-poor areas, or superimposed passages in chert-rich areas. In a vadose system (above the water table), the faults have several influences. First, they are the focus of collapse, increasing passage height, and in some cases facilitating the opening of the caves to the landscape. Second, they provide conduits for meteoric water recharge and localized dissolution. Allogenic sediment accumulation, erosion and speleothem growth are generally focused around these fault structures. The continued influence of these fracture systems is demonstrated by faulted calcified sediments and multiple generations of intrusive speleothem growth found in the vicinity of the joints and faults throughout the system.

The combination of tensional joint and fault systems and different chert densities has formed two main karst morphologies at Sterkfontein. An upper level is represented by a single, large, deep chamber named the Fossil Cavern. The infilling deposits are now exposed on the surface by a combination of landscape erosion and then lime mining activities removing the roof of the chamber (Fig. 17.4). The lateral dimensions of the cavern are controlled by the convergence of five faults (Fig. 17.3). Enormous collapsed blocks (>5 × 5 × 5 m) at the base of the Fossil Cavern excavation suggest that the chamber has expanded through vadose collapse through its evolution, possibly activating different openings and depositing sediments in different areas.

The lower level of the cave system is more extensive than the upper, with passage lengths in excess of 4.5 km. Here, passage development is influenced by fault direction, fault density and chert-poor dolomite, creating very tall, narrow passages. Passages are generally labyrinthine to the east and parallel to the west, developing closely along active compound tensional fracture systems that are slightly more distant from one another (Gallery A and B in Fig. 17.3). The density of fracture systems exerts an important control on the lateral dimensions of these chambers and passages, with vadose collapses joining spaces between tall, narrow, closely parallel passages, forming galleries (e.g. 'Gallery B' in the Milner Hall; Fig. 17.5). Inter-gallery passages are generally smaller and perpendicularly orientated, developing along the subordinate north–south trending faults.

17.5 Cave Deposits

Opening of the caves to the surface occurs through a combination of vadose collapse and localized meteoric dissolution along faults. Openings are often very deep and, in some cases, connect the landscape directly to the base of the

Fig. 17.4 The Sterkfontein excavation site. The red allogenic sediments are visible against the grey dolomite. On the hill slope facing in the distance is the site of Swartkrans. (*Photograph* D. Stratford)

Fig. 17.5 Sterkfontein Milner Hall passage wall remnants. Note evidence of the vadose collapse on the cave floor, with the dolomite and interbedded chert dipping to the north (right). (*Photograph* D. Stratford)

vadose system about 30 m below (Fig. 17.6). Where surface erosion has intersected with the uppermost passages, long deep gullies are opened. Typically, openings are very steep sided and represent a serious danger to people and wild animals. Consequently, 'death trap' assemblages represent a common bone accumulation agent. Concentration of vegetation around the cave openings, particularly the wild olive (*Olea oleaster*), is thought to have provided shelter for primates, carnivores and sometimes hominins.

Once open, the caves start to accumulate allogenic sediments, animal bones, stone tools and occasionally the bones of our hominin ancestors. Sediments generally accumulate through colluvial processes with varying degrees of water interaction, developing a range of sedimentary facies related to talus cone formation. Within the cave, sediments can calcify, decalcify, collapse, and be eroded and reworked by meteoric and percolating water. These processes spread sediments, and their interned bones and artefacts, vertically and laterally en masse or as separate components through the system, sometimes stratigraphically isolating assemblages from their sources. Many of the deposits found at Sterkfontein are heavily calcified which, although providing a means of preserving fossils in situ for millions of years, require specialist excavation methods that are either small scale and very time-consuming, or heavy duty and problematic for stratigraphic control.

As the surface is eroded, new openings to the deepest chambers have been formed, and old openings have been opened again, depositing new sediments across the system and at all levels. Where erosive water is introduced, old deposits have been undercut and voids filled with new material. In some cases, re-karstification within autogenic and allogenic cave deposits has created irregular and vertical unconformities between sediment bodies. Such a situation can be seen in newly exposed deposit surfaces in the western area of the Fossil Cavern. The combination of these processes creates a highly dynamic depositional environment with complex stratigraphic histories.

Currently, the deposits found within the Fossil Cavern and its lower reaches (Silberberg Grotto) are organized into a macro-scale chronostratigraphic sequence (consecutively

Fig. 17.6 A typical cave opening surrounded by established vegetation at Sterkfontein. This particular opening articulates directly to the base of the system some 25 m below. (*Photograph* D. Stratford)

formed deposits classified by relative age) of six members (M1-6), referred to as the 'Sterkfontein Formation' (Partridge and Watt 1991). Of these, the most noteworthy are Member 2, the oldest allogenic deposit identified (between 4 and 2.8 Ma; Pickering and Kramers 2010) which contained the world's most complete single *Australopithecus* skeleton—'Little Foot', and Member 4, the richest *Australopithecus*-bearing deposit in the world, which has yielded over 750 specimens.

17.6 Summary

The geomorphology of the Cradle of Humankind documents a long and complex history of landscape and environmental processes spanning over two billion years since the deposition of dolomitic limestone, which hosts the abundant caves. This is highlighted with a flash of hominin evolutionary history documented over the last three million years and which is preserved within fossiliferous deposits of the Sterkfontein Caves. The cumulative influence of those processes has developed a unique sedimentary context in which we find the fossils of our hominin ancestors, their early technology and coexisting fauna. The preservation of these assemblages is largely owing to the fortuitous opening of the caves about three million years ago, facilitating the documentation of landscape and hominin evolution over the last several million years—a period of our evolution that is so intriguing due to the high levels of morphological variation found within the many different species of Plio-Pleistocene hominins inhabiting the southern African landscape. The varying lithologies and faulting of the dolomite have influenced the morphology of the cave network, the distribution of flora on the landscape, when and how the cave opened, and how sediments, bones and artefacts were deposited and distributed around the cave. The result is a dynamic karst network with a long and complicated depositional history. The Sterkfontein Caves and the Cradle of Humankind was awarded UNESCO World Heritage status in 1999 and is recognized as an invaluable cultural asset to South Africa. The Sterkfontein Caves are currently the only publicly accessible cave in the Cradle area and are visited daily by schools, tourists and students.

References

Bruxelles L, Quinif Y, Wienin M (2009) How can ghost rocks help in karst development? In: Proceedings of 15th international congress of speleology, Speleogenesis, pp 814–818

Catuneanu O, Eriksson PG (2002) Sequence stratigraphy of the Precambrian Rooihoogte—Timeball Hill rift succession, Transvaal Basin, South Africa. Sed Geol 147:71–99

Clarke RJ (2013) *Australopithecus* from Sterkfontein Caves, South Africa. In: Reed KE, Fleagle JG, Leakey RE (eds) The palaeobiology of *Australopithecus*, vertebrate palaeobiology and palaeoanthropology. Springer, Dordrecht, pp 105–123

Dirks PHGM, Berger LR (2013) Hominin-bearing caves and landscape dynamics in the Cradle of Humankind, South Africa. J Afr Earth Sci 78:109–131

Dubois C, Quinif Y, Baele J-M, Barriquand L, Bini A, Bruxelles L, Dandurand G, Havron C, Kauffmann O, Lans B, Maire R, Martin J, Rodet J, Rowberry MD, Tognini P, Vergari A (2014) The process of ghost-rock karstification and its role in the formation of cave systems. Earth Sci Rev 131:116–148

Eriksson PG, Hattingh PJ, Altermann W (1995) An overview of the geology of the Transvaal sequence and bushveld complex, South Africa. Miner Deposita 30:98–111

Herries AIR, Shaw J (2011) Palaeomagnetic analysis of the Sterkfontein palaeocave deposits: Implications for the age of the hominin fossils and stone tool industries. J Hum Evol 60:523–539

Klimchouk AB (2007) Hypogene Speleogenesis: Hydrogeological and Morphogenetic Perspective. Special Paper no. 1, National Cave and Karst Research Institute, Carlsbad, NM, 106 pp

Martini JEJ, Wilinger PE, Moen HFG, Keyser A (2003) Contribution to the speleology of Sterkfontein cave, Gauteng province, South Africa. Int J Speleol 32:43–49

Moen HFG, Martini JEJ (1996) The exploration of Swartgat Sinkhole. S Afr Spelaeological Assoc Bull 36:45–49

Partridge TC, Watt IB (1991) The stratigraphy of the Sterkfontein hominid deposit and its relationship to the underground cave system. Palaeontol Africana 28:35–40

Pickering R, Kramers JD (2010) Re-appraisal of the stratigraphy and determination of new U-Pb dates for the Sterkfontein hominin site, South Africa. J Hum Evol 59:70–86

Stratford DJ, Grab S, Pickering TR (2014) The stratigraphy and formation history of fossil- and artefact-bearing sediments in the Milner Hall of Sterkfontein Cave, South Africa: new interpretations and implications for palaeoanthropology and archaeology. J Afr Earth Sci 96:155–167

Sumner DY, Grotzinger JP (2004) Implications for Neoarchaean ocean chemistry from primary carbonate minerology of the Cambellrand-Malmani platform, South Africa. Sedimentology 51:1273–1299

Truswell JF, Eriksson KA (1975) A palaeoenvironmental interpretation of the early proterozoic Malmani Dolomite from Zwartkops, South Africa. Precambr Res 2:277–303

Wilkinson MJ (1973) Sterkfontein cave system: evolution of a karst form. M.A. thesis, University of the Witwatersrand, Johannesburg

Wilkinson MJ (1983) Geomorphic perspectives on the Sterkfontein Australopithecine breccias. J Archaeol Sci 10:515–529

Military Geography: Landscapes of the Anglo-Boer War

18

David G. Passmore, Jasper Knight, and Stephan Harrison

Abstract

The Anglo-Boer War of 1899–1902 was a significant conflict in the recent history of South Africa, but the military geography of this conflict has not been subject to systematic analysis. This chapter explores the relationship between the strategies of military engagement during this conflict and the nature of the physical landscape in which these engagements took place. An overview of the broad types of geomorphological and geological settings for some 43 major engagements of the conflict is followed by a detailed case study of the landscape contexts of engagements around Colenso (former province of Natal). The relatively limited degree of post-conflict land use and geomorphological change in these parts of South Africa also renders many of these battlefield sites readily appreciated in the landscape and is reflected in an increasing interest in battlefield heritage tourism.

Keywords

Battlefields • Anglo-Boer War • Koppies • Military geography • Colenso

18.1 Introduction

The Anglo-Boer War, fought between the independent Boer Republics and Great Britain between October 1899 and May 1902, has the distinction of being the longest, costliest and bloodiest military campaign fought by the British in the hundred years before the outbreak of World War I and is widely recognised for its major political and social repercussions both within South Africa and also in Britain (Pakenham 1979; Cuthbertson and Jeeves 1999; Judd and Surridge 2013). It is also understood that the conflict was militarily significant in that it presented the first occasion in which armed forces had engaged each other using recognisably modern weaponry, including machine guns, magazine-fed rifles and high-explosive artillery shells, whilst also presaging the effectiveness of guerilla-type warfare (Wessels 2011).

Although many of the Anglo-Boer War battlefields are well known and described (e.g. Marix Evans 1999), there have been few attempts to synthesise the military geography of the conflict and especially the influence of landscape geomorphology on the development of military tactics during the conflict and the nature and outcomes of military engagements. These are longstanding issues of interest to military geographers (e.g. Winters et al. 1998; Woodward 2014) and are explored here in the context of the Anglo-Boer War with the aim of providing (i) an overview of military engagements in the conflict with respect to their geographical locations and landscape context and (ii) a more detailed case study of battles in the Tugela River valley at Colenso, KwaZulu-Natal (formerly Natal province), which illustrates the interplay between the physical geography of the terrain and the military tactics used during engagement. Finally, some consideration is given to the value of former battlefield sites and their geological and geomorphological context in

D.G. Passmore (✉)
University of Toronto Mississauga, Mississauga, Canada
e-mail: david.passmore@utoronto.ca

J. Knight
University of the Witwatersrand, Johannesburg, South Africa

S. Harrison
University of Exeter, Exeter, UK

S. Grab and J. Knight (eds.), *Landscapes and Landforms of South Africa*,
World Geomorphological Landscapes, DOI 10.1007/978-3-319-03560-4_18,

terms of heritage management and archaeological preservation. The locations of battles described herein are given using their former province names (Table 18.1).

18.2 Military Geography of the Anglo-Boer War

Military actions in the Anglo-Boer War ranged from relatively minor skirmishes, including those typical of guerilla-type engagements, to larger battles (involving thousands of combatants) that could last for several days or, in the case of the well-known sieges of Ladysmith, Mafeking and Kimberley, for several months. In this study, we have considered the geographical context of 43 discrete skirmishes and battles that are identified and described by Pakenham (1979) and Judd and Surridge (2013), augmented by reference to papers that deal specifically with the different battles (e.g. Harris 1994). Table 18.1 lists the dates and locations of these engagements together with their outcomes and topographic context; engagement locations are mapped in Fig. 18.1 which also shows the major physiographic regions in the area of conflict (identified after Partridge et al. 2010), contemporary borders and major contemporary railway lines.

The regional geography of military engagements of the Anglo-Boer War reflects three broad phases. Beginning in October 1899, the first phase saw Boer incursions into the Cape Colony and Natal territories and besieging of British garrisons at Kimberley, Mafeking (both in the Cape Colony) and Ladysmith (Natal; Fig. 18.1). Accordingly, many of the early engagements of the war occurred in the British colonial territories, especially Natal (13 engagements; Table 18.1), and many of these constituted humiliating defeats for the British forces striving to relieve the sieges, most notably at Magersfontein and Colenso (discussed below). In the second phase, commencing early in 1900, heavily reinforced British forces pushed north to lift the sieges, and these forces advanced into the Boer Republics of the Transvaal and Orange Free State, eventually taking Pretoria, the Transvaal's capital, in June 1900. Over half of the engagements considered here were fought in these territories (22 engagements; Table 18.1). An extended period of guerilla raiding by Boer groups targeting British supply columns, depots and communications infrastructure characterised military actions during the third phase of the war which extended from mid-1900 to the end of the war in May 1902.

British lines of advance during the first and second phases of the war were strongly influenced by the dependence of British forces on railways for transport and logistical support (Wolmar 2010). This accounts for the linear pattern of the location and sequencing of many engagements, especially in the drives towards Ladysmith and Kimberley (Fig. 18.1). At the scale of the battlefield, Pakenham (1979) notes that the deployment of long-range, magazine-fed rifles was to greatly extend the space over which infantry could engage and, when combined with smokeless cartridges and entrenched firing positions, generally acted to tilt the battlefield advantage in favour of the defender (and in this respect presage the battlefields of the forthcoming World War I). With the exception of siege locales, the immediate location and terrain context of engagements were typically forced by Boer defensive stands and ambushes, and here, the Boers often proved adept at exploiting the physical geography (terrain) features to suit their tactics and approach to warfare. Table 18.1 illustrates that Anglo-Boer War battlefields can be grouped into three broad categories of terrain, and these are described as follows. Figure 18.2 illustrates the range of terrain contexts encountered at the battle site of Colenso (Natal), which is described in detail below.

18.2.1 Kopjes and Ridges

Much of the Orange Free State and southern Transvaal regions occupy the interior plateau of the Highveld, a Late Cretaceous erosion surface largely developed on Permian (Ecca and Beaufort Groups) and Triassic (Beaufort Group) near-horizontal shales, mudstones and sandstones (Partridge et al. 2010). These gently undulating grassland landscapes (locally termed 'grassveld') offered little obstacle to the manoeuvre of British troops and their attendant artillery and supply columns. Many areas, for example in large parts of the western Free State, saw little or no combat (Amery 1906). In seeking suitable terrain for ambush and defensive stands, the Boer forces chose instead to exploit locales where the low relief of the Highveld is intruded by outcrops of relatively resistant geologies, especially dolomite, granite and quartzite, giving rise to irregular rocky hills and ridges that rise up to several tens of metres above the local surfaces (Fig. 18.2a). These are generally termed kopjes (koppies) but may include larger monadnocks and inselbergs.

Even relatively small changes in elevation may confer tactical advantages to military forces in relatively open landscapes (e.g. Winters et al. 1998), and the Boers were adept at exploiting the enhanced fields of view and good cover for gun and rifle positions afforded by kopjes. Indeed, kopjes and ridges account for 56 % of the engagement locations identified in this study, including the majority of early battles in 1899 and the famous action at Spion Kop early in 1900 (Table 18.1). However, the rocky summits of

Table 18.1 List of dates and locations of engagements in the Anglo-Boer War together with their topographic context; see Fig. 18.1 for map of locations

#	Date	Battle	Location (Province)	Topography	Attack
1	20 Oct 1899	Talana Hill	Dundee (Natal)	Kopje/ridge	British frontal assault
2	21 Oct 1899	Elandslaagte	Elandslaagte (Natal)	Kopje/ridge	British frontal assault
3	30 Oct 1899	Pepworth	Ladysmith (Natal)	Kopje/ridge	British frontal assault
4	23 Nov 1899	Belmont	Belmont (Cape Colony)	Kopje/ridge	British frontal assault
5	25 Nov 1899	Graspan	Graspan (Cape Colony)	Kopje/ridge	British frontal assault
6	28 Nov 1899	Modder River	Modder River (Cape Colony)	River crossing/valley floor	British frontal/flank assault
7	10 Dec 1899	Magersfontein	Magersfontein (Cape Colony)	Kopje/ridge	British frontal assault
8	10 Dec 1899	Stormberg	Stormberg Junction (Cape Colony)	Kopje/ridge	British frontal assault
9	15 Dec 1899	Colenso	Colenso (Natal)	River crossing/valley floor	British frontal assault
10	6 Jan 1900	Wagon Hill/Caesar's Camp (Platrand)	Ladysmith (Natal)	Kopje/ridge	Boer night assault
11	24 Jan 1900	Spion Kop	Spion Kop (Natal)	Kopje/ridge	British night assault
12	5 Feb 1900	Vaal Krantz	Vaal Krantz (Natal)	Kopje/ridge	British frontal assault
13	18 Feb 1900	Paardeberg	Paardeberg Drift (Orange Free State)	River crossing/valley floor	British frontal/flank assault
14	14–19 Feb 1900	Hussar Hill/Cingolo/Monte Cristo/Hlangwane Hill	Colenso (Natal)	Kopje/ridge	British flank assault
15	23–27 Feb 1900	Hart's Hill/Railway Hill/Pieters Hill	Colenso (Natal)	Kopje/ridge	British frontal/flank assault
16	7 Mar 1900	Poplar Grove	Poplar Grove (Orange Free State)	Kopje/ridge	British frontal/flank assault
17	31 Mar 1900	Sannah's Post	Sannah's Post (Orange Free State)	River crossing/valley floor	Boer ambush
18	30 Apr 1900	Houtnek	Houtnek Poorte (Orange Free State)	Valley/mountain pass	British frontal/flank assault
19	4 May 1900	Brandfort	Brandfort (Orange Free State)	Kopje/ridge	British frontal/flank assault
20	12 May 1900	Mafeking	Mafeking (Cape Colony)	Urban/building	Boer infiltration
21	10–15 May 1900	Biggarsberg	Dundee (Natal)	Valley/mountain pass	British flank assault
22	26 May 1900	Doornkop	Krugersdorp (Transvaal)	Kopje/ridge	British frontal assault
23	31 May 1900	Lindley	Lindley (Orange Free State)	Kopje/ridge	Boer frontal assault
24	11–12 June 1900	Diamond Hill (Donkerhoek)	Pretoria (Transvaal)	Kopje/ridge	British frontal assault
25	6 June 1900	Roodewal	Koppies (Orange Free State)	Urban/building	Boer frontal assault
26	6–12 June 1900	Allemans Nek/Laings Nek	Majuba (Natal)	Valley/mountain pass	British flank assault
27	11 July 1900	Zilikats Nek	Utival (Transvaal)	Valley/mountain pass	Boer frontal assault
28	23 July 1900	Slabberts Nek/Retiefs Nek	Fouriesburg (Orange Free State)	Valley/mountain pass	British frontal assault
29	27 Aug 1900	Belfast (Bergendal)	Belfast (Transvaal)	Kopje/ridge	British frontal assault
30	6 Nov 1900	Bothaville	Bothaville (Orange Free State)	River crossing/valley floor	British ambush
31	13 Dec 1900	Nooitgedacht	Nooitgedacht (Transvaal)	Valley/mountain pass	Boer ambush
32	29 May 1901	Vlakfontain	Naawpoort (Transvaal)	Kopje/ridge	Boer frontal assault
33	3 July 1901	Vlakfontain	Elandskloof (Transvaal)	Valley/mountain pass	Boer frontal assault

(continued)

Table 18.1 (continued)

#	Date	Battle	Location (Province)	Topography	Attack
34	5 Sept 1901	Groenkloof	Petersburg (Cape Colony)	Valley/mountain pass	British frontal assault
35	17 Sept 1901	Elands River	Elands River Poort (Cape Colony)	Valley/mountain pass	Boer frontal assault
36	17 Sept 1901	Blood River Poort (Scheeper's Nek)	Blood River Poort (Natal)	Valley/mountain pass	Boer flank assault
37	30 Oct 1901	Bakenlaagte (Gun Hill)	Bakenlaagte (Transvaal)	Kopje/ridge	Boer frontal assault
38	25 Dec 1901	Groenkop (Tweefontain)	Tweefontain (Orange Free State)	Kopje/ridge	Boer frontal assault
39	26 Dec 1901	Fort Itala/Fort Prospect	Nkandhla (Natal)	Kopje/ridge	Boer frontal assault
40	24 Feb 1902	Yzer Spruit	Orkney (Transvaal)	River crossing/valley floor	Boer ambush
41	7 Mar 1902	Tweebosch (De Klipdrift)	Tweebosch (Transvaal)	River crossing/valley floor	Boer ambush
42	31 Mar 1902	Boschbult	Boschbult (Transvaal)	Kopje/ridge	Boer ambush
43	11 Apr 1902	Rooiwal	Rooiwal (Transvaal)	Kopje/ridge	Boer frontal assault

kopjes were not impregnable and, particularly if isolated from supporting forces, could be overwhelmed by a combination of British artillery fire and determined assault, even if the latter were often costly affairs fought over steep, rocky slopes covered with scrub and sometimes cut by gullies. For example, at the Battle of Bergendal (eastern Transvaal, Table 18.1; Fig. 18.1), fought on 27 August 1900, and one of the last major battles of the war, the Boer position was made vulnerable when the defenders of an isolated kopje in front of the main line of defence were overrun despite a valiant stand described by the historian Conan Doyle (1902) as one of the finest defensive actions of the war; here, the decisive role was played by British artillery firing from concealed positions well beyond the range of Boer guns (Pakenham 1979; Jooste 2002).

Accordingly, the defensive advantages conferred by kopjes were often most effective where hilltops clustered close together giving mutually supporting vantage points (see for example, the actions at Colenso; Fig. 18.2a–c). Similarly, elongate ridges also offered the advantage of height whilst serving as barriers that were difficult to outflank or which channelled attacking forces into narrow valleys and passes. On other occasions, however, the Boers had proved adaptable to the vulnerability of isolated kopje positions to artillery fire. Following casualties sustained to shrapnel in hilltop stands at Graspan and Belmont in November 1899 (Cape Colony, Table 18.1; Fig. 18.1), Boer forces at the subsequent battle at Magersfontein entrenched in well-concealed positions along the lower footslopes of the prominent kopje and were spared the preliminary British artillery fire that was aimed at assumed hilltop positions (Davitt 1902; Pakenham 1979).

18.2.2 River Crossings and Adjacent Valley Floors

It has long been recognised that rivers present potentially formidable obstacles to advancing forces whilst being advantageous to defenders (Winters et al. 1998) and this proved to be the case on at least seven occasions during the Anglo-Boer War (Table 18.1). Many rivers in the South African interior exhibit a mixed alluvial-bedrock character with the latter being commonly associated with outcropping harder bedrock types (Tooth et al. 2007). As a consequence, such rivers tend to exhibit little morphological variability over decadal to centennial timescales, compared to alluvial rivers, which means that their condition today is likely very similar to that during the conflict. Most permanent (bridged) river crossings associated with contemporary roads and railway lines in the interior were located in relatively wide alluvial plains and channels in these locations were (and remain) deeply incised into alluvial deposits and underlying bedrock with steep and thickly vegetated banks (Fig. 18.2b). Small dongas or gullies are often cut into river banks and adjacent alluvial surfaces and have been recognised as offering local cover and concealment to combatants (Fig. 18.2d); but the presence of bridges or fording points (drifts) more typically acted as the key controls on the location and pattern of engagements. The presence of drifts assumed particular importance where bridges were destroyed

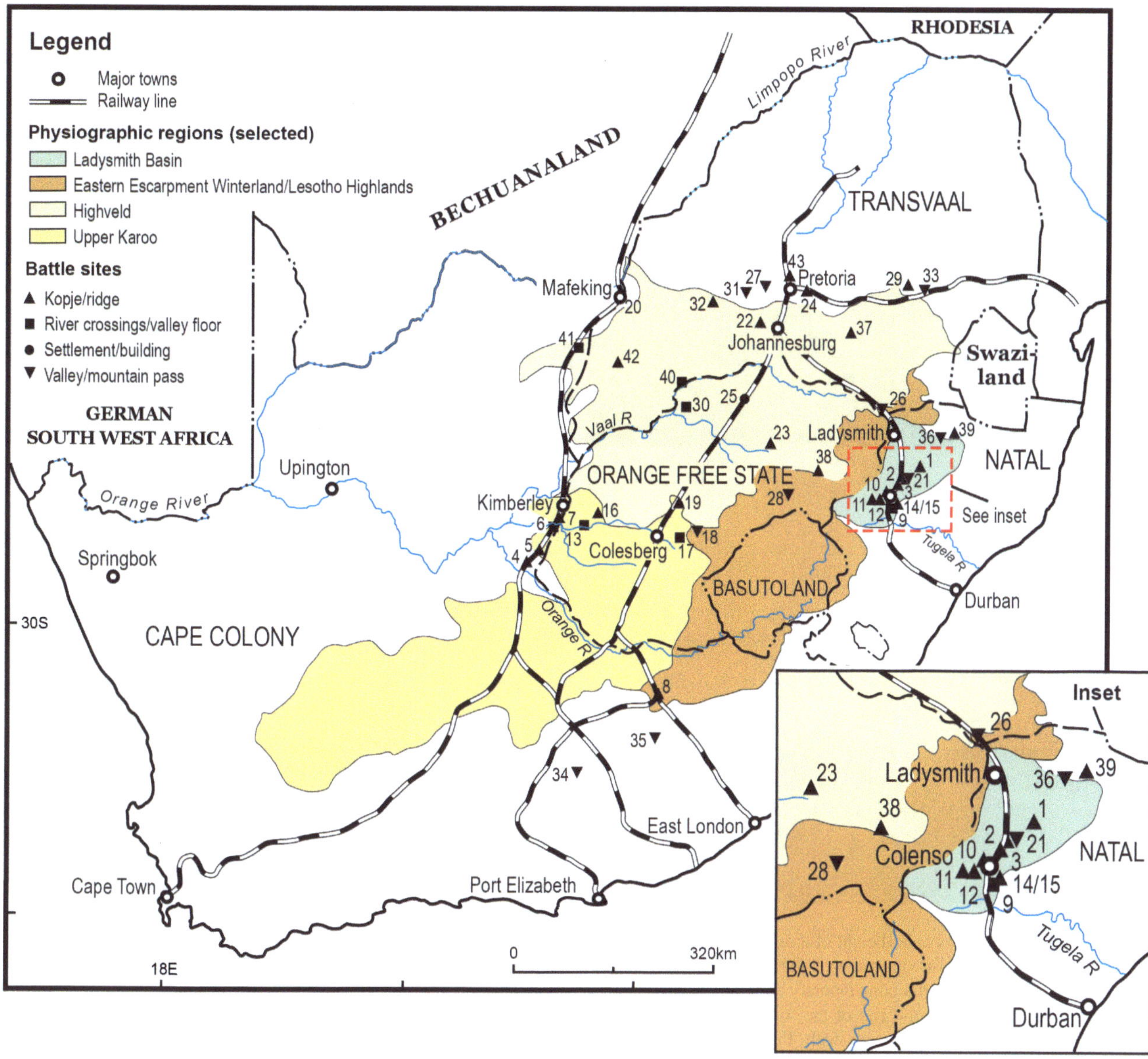

Fig. 18.1 Map of southern Africa showing Colony and Republic borders at 1899, major physiographic regions mentioned in the text, principal railway lines and locations of engagements over the period 1899–1902. For key to engagements see Table 18.1

but, whilst their locations were well known to Boer commanders, they were rarely depicted on the poor-quality maps available to the British (Marix Evans 2002). Accordingly, attacking British columns were frequently required to scout crossing points and improvise tactics during actual assaults, and on at least one occasion this lack of local knowledge led to a major setback in the battle.

18.2.3 Valleys and Mountain Passes

In areas of relatively high relief, notably in the elongated quartzite ridges of the Pretoria Group in the western Transvaal and in the Eastern Escarpment hinterland and Drakensberg highlands, relatively confined valleys and mountain passes were exploited as ambush sites. On 31 March 1990 at Sannah's Post (Orange Free State; Table 18.1; Fig. 18.1), for example, Boer riflemen located on hilltops drove British forces into a river valley where they were captured by waiting Boer troops. Later that year on 11 June at Zilikats Nek, a pass cut through a quartzite escarpment in the Magaliesberg region of the western Transvaal (Fig. 18.1; Table 18.1), British troops in a poorly prepared bivouac were humiliated in a Boer ambush. Here, the relief in the Nek was such that the British artillerymen were unable to elevate their guns high enough (even having excavated beneath the gun trails) to engage the Boers occupying positions high on the overlooking quartzite ridge (Copley 1993).

Fig. 18.2 **a** Example of a kopje (height of around 15 m) located 4 km north of Colenso; **b** downstream view of the Tugela River, located immediately downstream of the contemporary road bridge at Colenso. A modern weir is located at the site of the former Wagon Drift (see Fig. 18.3 for location and compare with Fig. 18.4); **c** viewpoint northwards from the position of Long's Guns towards the Tugela River. *Note* the hilltop vantage points in the distance; **d** typical river crossing point near Wynne Hill. Note the mixed gravel/bedrock substrate and the presence of rocky cliffs which would have afforded troop cover. (*Photographs* J. Knight)

18.3 Terrain and Tactics in the Battlefields Around Colenso (Natal)

Actions in the vicinity of Colenso occurred on 15 December 1899 and subsequently over 14–27 February 1900 and were associated with attempts by British forces under General Sir Redvers Buller to force a passage to besieged Ladysmith located 20 km to the north. These actions were among the largest battles of the Anglo-Boer War, involving upwards of 20,000 combatants, and both shared the requirement of crossing the Tugela River before driving a passage through a series of rugged kopjes on the northern side of the river that command a clear view over the alluvial plain below (Fig. 18.2b).

In common with most rivers draining east from the Drakensberg, the Tugela River exhibits distinctive lithological and structural controls on its channel and floodplain character and, in particular, is strongly controlled in its course by the Tugela fault. North of this fault line, and thus the Tugela River, impermeable shales dominate, which underlie the prominent kopjes. South of the river, interbedded mudstone and sandstone are present and the area has less prominent relief (Fig. 18.2c). In the Colenso area, these geologic controls are locally evident in the transition of the Tugela from a relatively high-sinuosity channel and wide, low-relief alluvial plain upstream of the town, to a relatively confined and narrow valley floor 6 km downstream of the town where the Tugela enters a steeply incised fault-controlled gorge with several bedrock islands and falls

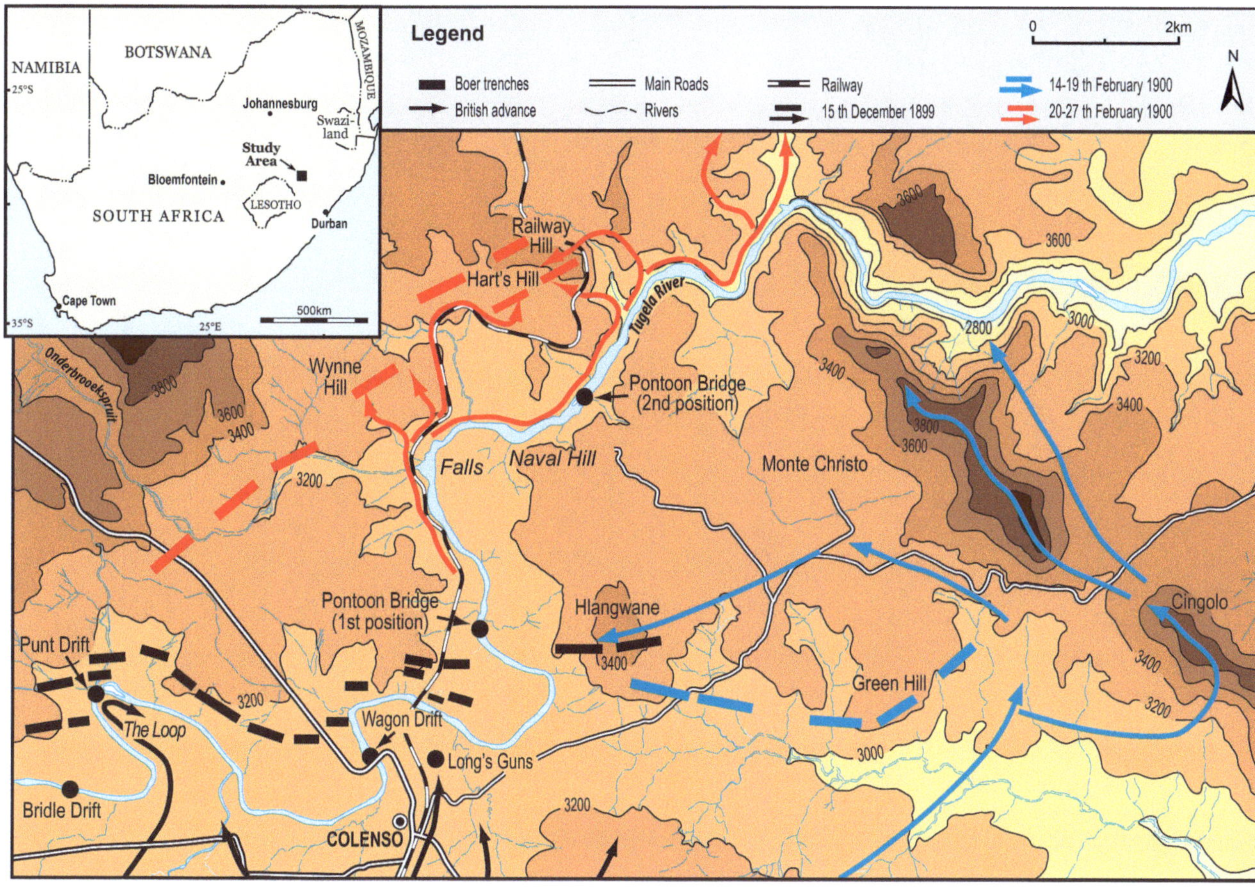

Fig. 18.3 Map of the Tugela River valley at Colenso showing relief, bridge and drift sites, locations of primary Boer defence positions and British attacks for the two battle periods of 15 December 1899 and 14–27 February 1900

(Fig. 18.3). The Tugela channel is in a long-term state of incision and has a channel bed lying close to or directly traversing bedrock (Tooth et al. 2004; Grenfell et al. 2008). Contemporary maps and accounts indicate little planform change over the period since the conflict and hence the present river environment gives a good impression of the obstacle confronting Buller, with an incised channel up to 150 m wide and steep vegetated banks between 4.5–6 m high (Fig. 18.2b).

Of the road and railway bridges spanning the Tugela at Colenso in 1899 (Fig. 18.4), only the former was intact at the outset of the engagements, and hence the availability of fording points, or 'drifts' assumed particular military importance. Located in shallower reaches of the channel, and often immediately upstream of bedrock island and bar complexes, the drifts were locally consolidated with emplaced rocks that had been wholly or partially removed by the Boers in preparation for the defence. However, the river does not appear to have been in flood and contemporary accounts describe water levels as 'breast-deep' and passable (Davitt 1902; Pakenham 1979). Nevertheless, drifts provided predictable foci for British attacks and hence were choke points that were covered by well-entrenched Boer forces.

On 15 December 1899, Buller launched his first and ultimately unsuccessful attack as a frontal assault in three columns, intended to secure the river crossings at Bridle Drift to the west, the intact road bridge and Wagon Drift in Colenso, respectively, and also the dominating relief of Hlangwane, a scrub and ravine-cut kopje dominating the south bank of the Tugela east of Colenso (Fig. 18.3). All three columns were repulsed by massed rifle fire with artillery support, aided by tactical errors by the western and central columns. In the absence of good-quality maps, the western column advanced in error on Punt Drift at the apex of a northerly meander bend of the Tugela which brought it under fire from three sides (Pakenham 1979; Marix Evans 2002). Meanwhile, matters in the central column were not helped by the advance of a gun battery (under Colonel Long) into an exposed forward position that was well within effective rifle range of the Boers on the north bank of the Tugela (Figs. 18.2c and 18.3). Attempts to extract this battery were only partially successful. To the east, the assault on

Fig. 18.4 1902 *photograph* of railway bridge crossing the Tugela River at Colenso, which was destroyed by Boer forces. The present town is on the lower right bank. *Source* Wikimedia Commons, available from http://commons.wikimedia.org/wiki/File:BoersPontChemin_de_fer_de_Colenso,_sur_laTugela,_d%C3%A9truit_parBoersVers1900.jpg. Compare this view with that today (see Fig. 18.2b)

Hlangwane also foundered under intense rifle fire and the difficulty of traversing the kopje slopes.

In his second attack at Colenso in February 1900, Buller adopted new tactics based around assaults covered by 'creeping barrages' that exploited the numerical advantage in artillery enjoyed by the British. In the first phase of the battle between 14 and 19 February, a series of kopjes to the east of Colenso were gained in turn, thereby securing the southern side of the Tugela and achieving elevated positions able to cover a river crossing via a pontoon bridge to the west of Hlangwane (Fig. 18.3). Once established on the northern side of the Tugela, the British commenced a series of assaults on the hills barring the route to Ladysmith, these being forced with the help of a flanking operation via a second pontoon bridge downriver of Naval Hill.

18.4 Military Geoheritage and Tourism

The former battlefields of the Anglo-Boer War are becoming increasingly familiar as destinations for visitors and professionals with an interest in military history. In the KwaZulu-Natal region, for example, many of the sites associated with Buller's campaign to relieve Ladysmith are now established in visitor guides and tours alongside the famous Anglo-Zulu War battlefield sites of Isandlwana and Rorke's Drift, and there is a dedicated Siege Museum at Ladysmith. Several Anglo-Boer War battlefield sites or the encampment areas associated with troop movements are also recognised as Provincial Heritage Sites, which can serve a dual purpose for the protection of both military heritage and landscape geomorphology.

Widespread memorials (especially to British troops) present prominent testimony to battlefield fatalities (Fig. 18.5) and some battlefield sites such as at Colenso are also provided with location markers and/or information boards (Fig. 18.6). Many of these memorials are at or near to the battlefields, thus are commonly on hilltop locations. At a memorial garden outside of Colenso at Clouston, pillars and plaques commemorate British officers killed in the conflict, and a number of memorials found here were moved to this site in the 1960s as the town of Colenso was developed. A visitor's book is also held here; over half the entries explicitly cite battlefield tourism as a motivation for visitors. In a wider South African context, the remembrance and memorialisation of past conflicts, from different perspectives, is an important present issue of cultural heritage management (e.g. Meskell and Scheermeyer 2008).

Fig. 18.5 Memorial at Wynne Hill to the 2nd Battalion, Prince Albert's Somerset Light Infantry. This memorial overlooks the pontoon bridge crossing point on the Tugela River (background) that was used during the second battle, 14–27 February 1900 (*Photograph* J. Knight)

Fig. 18.6 Information markers for gun positions associated with Colonel Long's battery in the Battle of Colenso, 15 December 1899 (*Photographs* J. Knight)

18.5 Conclusions and a Military Geography Agenda

Much has been written on the context, conduct and consequences of the Anglo-Boer War, and a growing list of battlefield tourism websites and tour guides testifies to its enduring interest. The geological, geomorphological and vegetation history of the South African interior landscapes have made a perhaps under-appreciated contribution to this battlefield history. Whilst few of the terrain features described here should have presented an insurmountable obstacle to attacking forces, and by the early–mid-1900s had in many cases been accommodated by revised British tactics, it can be argued that the combination of terrain, massed firepower and concealment were used with great effect by Boer forces and typically resulted in a disproportionate number of British casualties even if they did succeed in their ultimate objectives.

Modern appreciation and study of these battlefield sites is also facilitated by the relatively limited degree of post-conflict land use and geomorphological change (including river channel environments) in these parts of South Africa. Although infilled, evidence of trench positions can still be seen north of the Tugela River, for example, and in this respect the military geography and archaeology of the Anglo-Boer War may be more accessible by comparison with many recent historic conflict landscapes that have been impacted by subsequent urban expansion and agricultural intensification (see for example Passmore et al. 2014). There is, therefore, considerable potential for more detailed evaluation of the geoarchaeology of battlefield sites, especially through the demonstrable utility of LiDAR-based high-resolution topographic mapping and geomorphological analysis of military infrastructure (e.g. Maio et al. 2013).

References

Amery LS (ed) (1906) The Times History of the War in South Africa: 1899–1902, vol. 4. Sampson Low, London, 597 pp

Doyle AC (1902) The Great Boer War. Nelson & Sons, London

Copley IB (1993) The Battle of Silkaatsnek—11th July 1900. S Afr Mil Hist Soc Mil Hist J 9(3)

Cuthbertson G, Jeeves A (1999) The many-sided struggle for Southern Africa, 1899–1902. S Afr Hist J 41:2–21

Davitt M (1902) The Boer Fight for Freedom. Funk & Wagnalls Company, New York, p 603

Grenfell MC, Ellery WN, Grenfell SE (2008) Tributary valley impoundment by trunk river floodplain development: a case study from the KwaZulu-Natal Drakensberg Foothills, eastern South Africa. Earth Surf Proc Land 33:2029–2044

Harris JD (1994) The other Vlakfontein: an action in the Anglo-Boer War, 3 July 1901. S Afr Mil Hist Soc Mil Hist J 9(6)

Jooste CP (2002) The battle of Bergendal: the last pitched battle of the Anglo-Boer war. S Afr Mil Hist Soc Mil Hist J 12(4)

Judd D, Surridge K (2013) The Boer War: a History. IB Tauris & Co, London

Maio CV, Tenenbaum DE, Brown CJ, Mastone VT, Gontz AM (2013) Application of geographic information technologies to historical landscape reconstruction and military terrain analysis of an American Revolution Battlefield: preservation potential of historic lands in urbanized settings, Boston, Massachusetts, USA. J Cult Heritage 14:317–331

Marix Evans M (1999) The Boer War: South Africa 1899-1902. Osprey, Oxford 160 pp

Marix Evans M (2002) Maps and decisions: Buller South and North of the Tugela, 1899–1900. In: Doyle P, Bennett MR (eds) Fields of Battle: Terrain in Military History. Kluwer Academic, Dordrecht, p 387

Meskell L, Scheermeyer C (2008) Heritage as therapy—set pieces from the new South Africa. J Mat Cult 13:153–173

Pakenham T (1979) The Boer War. Weidenfeld and Nicholson, London

Partridge TC, Dollar ESJ, Moolman J, Dollar LH (2010) The geomorphic provinces of South Africa, Lesotho and Swaziland: a physiographic subdivision for earth and environmental scientists. Trans R Soc S Afr 65:1–47

Passmore DG, Harrison S, Tunwell DC (2014) Second World War conflict archaeology in the forests of northeast Europe. Antiquity 88:1275–1290

Tooth S, Brandt D, Hancox PJ, McCarthy TS (2004) Geological controls on alluvial river behaviour: a comparative study of three rivers on the South African highveld. J Afr Earth Sci 38:79–97

Tooth S, Rodnight H, Duller GAT, McCarthy TS, Marren PM, Brandt D (2007) Chronology and controls of avulsion along a mixed bedrock-alluvial river. Geol Soc Am Bull 119:452–461

Wessels A (2011) Boer guerilla and British counter-guerilla operations in South Africa, 1899 to 1902. Scientia Militaria, S Afr J Milit Stud 39:1–24

Winters HA, Galloway GE Jr, Reynolds WJ, Rhyne DW (1998) Battling the elements: weather and terrain in the conduct of war. John Hopkins University Press, Baltimore 317 pp

Wolmar C (2010) Engines of war: how wars were won and lost on the railways. Atlantic Books, London 310 pp

Woodward R (2014) Military landscapes: Agendas and approaches for future research. Prog Human Geog 38:40–61

Geoheritage and Geotourism in South Africa

19

Jasper Knight, Stefan Grab, and Amanda B. Esterhuysen

Abstract

The geologic and geomorphological diversity of South Africa, the spectacular landscapes that result, and their internationally important ecological and cultural/archaeological associations, all make the landscapes of South Africa prime sites for geoheritage and geotourism. To date, however, there has been little work to develop this potential. This chapter describes examples of sites where geoheritage and geotourism activities have been developed in South Africa, and how other scenic South African landscapes can enhance their geoheritage and geotourism potential.

Keywords

Geological conservation • Landscape • Heritage • Geomorphology • Tourism • Geoarchaeology

19.1 Introduction

As discussed throughout this book, there are close interrelationships in South Africa between its geology and geological history, geomorphology, landscape characteristics, hydrology, ecology, cultural heritage and environmental resources. Although this is true for most parts of the world, these interrelationships are particularly significant in South Africa because of its exceptional present-day biodiversity and endemism, and its palaeontological record of human evolution (Zunckel 2003; Reynolds et al. 2011). Despite this, very few studies have explicitly discussed the role of geology and geomorphology in large-scale landscape evolution, ecology and human culture. The varied reasons for this may include its lack of institutional and public participation in geology (Reimold 1999; Chirikure 2013), lack of data on important geological or geomorphological sites (Schutte and Booysen 2010), conflation of geological with other ecological and cultural heritage issues (Reimold 1999), and problems with relevant legislation and management (Scheermeyer 2005; Cairncross 2011).

Appreciation of the geological and/or geomorphological heritage, termed *geoheritage*, of landscapes and their relationship to wider ecological and cultural heritage issues is viewed by Dowling (2011) as a low-risk method of enhancing tourist income and nature conservation, particularly in developing countries. Tourist activities which focus on geology and landscape geomorphology are termed *geotourism*. The use of the prefix 'geo' has proliferated in recent years as a result of increasing appreciation of the relationships between geology and landscape, and due to an increase in demand for sustainable, ethical geotourism (see Brocx and Semeniuk 2007). In addition to geoheritage and geotourism, the following terms are also commonly used (definitions adapted after Gray 2013):

- *Geodiversity*, the range or richness of geological and/or geomorphological features, types and/or their attributes found within a given region. Geodiversity is the geologic equivalent of biodiversity;
- *Georesources*, the valuing from different perspectives of particular geological and/or geomorphological features or their attributes;

J. Knight (✉) · S. Grab (✉) · A.B. Esterhuysen (✉)
University of the Witwatersrand, Johannesburg, South Africa
e-mail: jasper.knight@wits.ac.za

S. Grab
e-mail: stefan.grab@wits.ac.za

A.B. Esterhuysen
e-mail: amanda.esterhuysen@wits.ac.za

S. Grab and J. Knight (eds.), *Landscapes and Landforms of South Africa*,
World Geomorphological Landscapes, DOI 10.1007/978-3-319-03560-4_19,

- *Geoconservation*, the conservation of geological and/or geomorphological features or their attributes;
- *Geotope*, the specific geologic and/or geomorphologic associations that are found within a specific spatial area. A geotope is the geologic equivalent of an ecotope.

Organisations such as the International Union for the Conservation of Nature and Natural Resources (IUCN), the International Union of Geological Sciences (IUGS) and UNESCO have a role in landscape conservation on different scales and for different purposes, many of which include a geologic element. For example, many World Heritage Sites (WHS), inscribed by UNESCO, have an explicit geologic component. Examples from South Africa include Vredefort Dome (inscribed 2005), and the Fossil Hominid Sites of Sterkfontein, Swartkrans, Kromdraai and Environs (South Africa) (1999), which is now known as Fossil Hominid Sites of South Africa (FHSSA, name changed in 2013). *Global geoparks* are large-scale landscapes where geoheritage and geotourism are linked together. Geoparks are identified in many areas of the world on the basis of the geological relationships to landscape evolution, scenic beauty, environmental resources and human activity/cultural heritage. Geoparks are managed locally and nationally, with an emphasis on sustainable development and geotourism, and supported on an informal and ad hoc basis by UNESCO. There are presently over 100 geoparks worldwide; none is in South Africa. The smallest scale of geoheritage management is that of the *geosite* (also termed a *geomorphosite* where the geosite has geomorphic expression), which may, for example, represent a specific outcrop, headland, coastal cliff or river reach, landform or fossil locality.

The use of these different terms is a potential cause of confusion. There are also problems in accurately mapping, delineating, inventorising, describing and interpreting geologic and geomorphic features and their properties. This information is required to evaluate the relative importance of different sites, the threats to site integrity (such as development, contamination, climate change and erosion) and to conserve the site or its attributes for the future (Erikstad 2013). Consequently, several studies have proposed quantitative ways of measuring geodiversity or the relative value of one site when compared to another (e.g. Pereira and Pereira 2010; Ruban 2010). This is especially problematic when integrating different types of geologic, geomorphic and environmental data, and in making interconnections between physical, ecological and cultural attributes (Gray et al. 2013).

The scenic landscapes of South Africa and their geologic and geomorphic features have great potential for sustainable geoheritage and geotourism which link together with other important ecological and cultural landscape attributes (Fig. 19.1). However, thus far, geoheritage and geotourism have not been widely considered in a South African context (Reimold et al. 2006; Schutte and Booysen 2010). Although there are several hundred Provincial Heritage Sites, these are primarily for buildings and monuments and only 15 are based solely on geomorphological value (i.e. waterfalls, glacial pavements and natural rock bridges) (Table 19.1). A further seven Provincial Heritage Sites are for geological phenomena such as rare rock types or interesting exposures (e.g. Cretaceous deposits at Mzamba Beach, Dumortierite near Kenhardt). Further sites, although declared Heritage sites based on their archaeological, palaeontological and cultural value, have additional strong landscape/geomorphological appeal (e.g. Sterkfontein Caves, Langebaanweg Quarry fossil site). Heritage Resources were formerly declared as protected sites by the National Monuments Council, which erected plaques to highlight a few sites of geological/geomorphological interest, such as the natural rock bridge in the Pilgrim's Rest District of Mpumalanga Province (Fig. 19.2; Table 19.1). Under the National Heritage Resources Act, number 25 of 1999, the National Monuments Council was replaced by the South African Heritage Resources Agency (SAHRA), which very occasionally continued to recognise sites of geological/geomorphological values (e.g. Vredefort Dome). More encouraging has been the inscription of several South African landscapes as WHS. The iSimangaliso Wetland Park, Cape Floral Region Protected Areas and Vredefort Dome were inscribed for their natural value and Maloti-Drakensberg Park for both its natural and cultural values (Table 19.1). Provincial and local park authorities have also taken some recent initiatives to provide information boards on regional geology, some of which have been placed adjacent to information on the cultural/archaeological and faunal heritage, such as in the Drakensberg Transfrontier Park. A further example is the rock garden at the Walter Sisulu National Botanical Garden in Johannesburg, which displays rock types and their relations to plants across South Africa (Fig. 19.3).

This chapter discusses the exciting potential in South Africa for geoheritage-based landscape conservation and management, and the related potential for geotourism. This chapter (1) outlines the importance of geoheritage in the South African landscape; (2) describes the present legislative and management framework for protection of landscapes and geoheritage resources; and (3) provides specific examples of geoheritage and geotourism activities in South Africa.

19.2 The Importance of Geoheritage in the South African Landscape

The relationship between geology, geomorphology, landscape evolution, and landscape ecology and cultural patterns in South Africa has been described in several recent large-scale studies (Bailey et al. 2011; Cotterill and de Wit 2011).

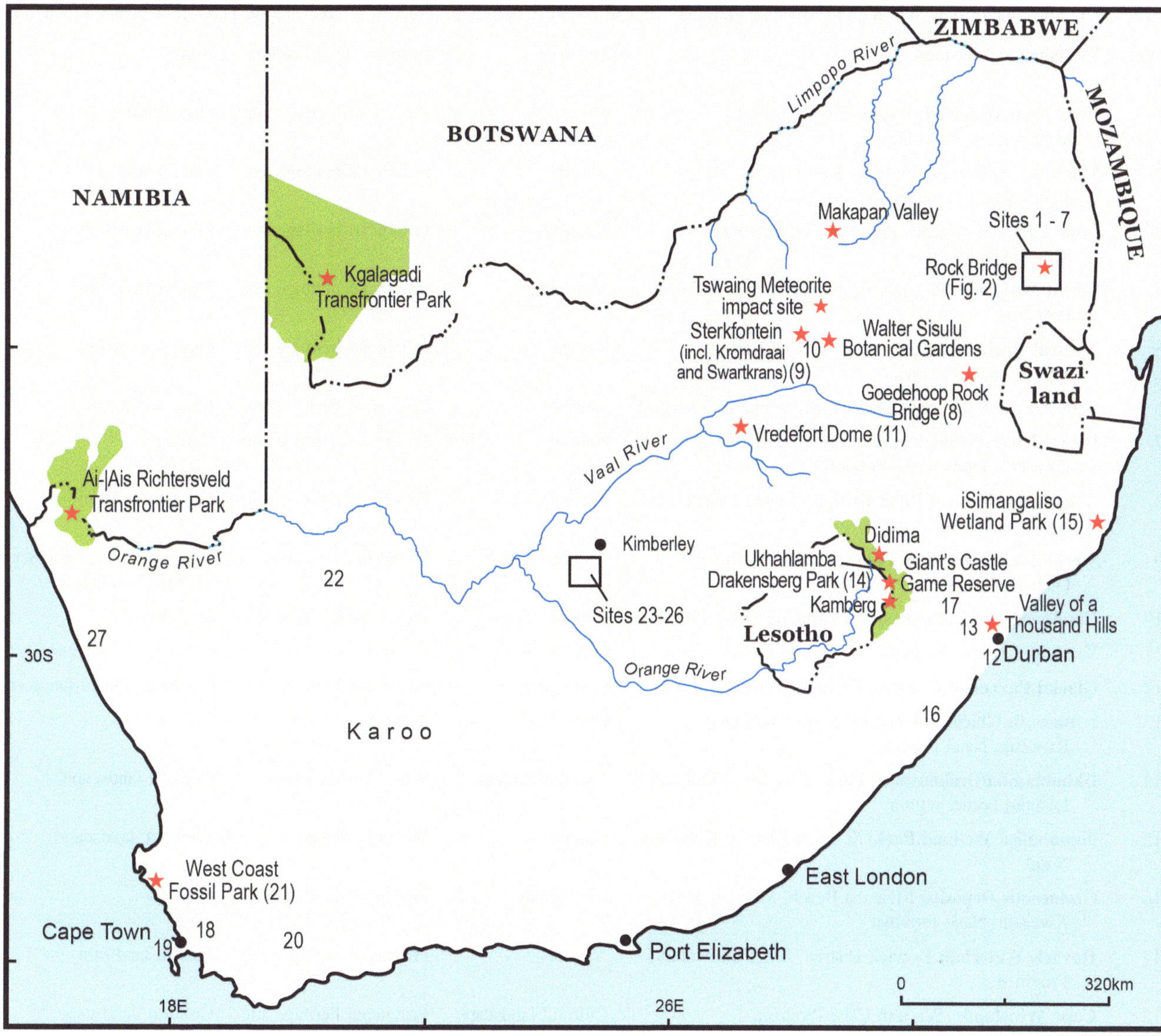

Fig. 19.1 Map of South Africa showing the locations of sites mentioned in the text

The problem with such studies is that landscape–culture relationships are often site-specific, manifested in different ways and change over time, largely as a result of changing patterns of human activity (Ruddiman 2012). Such viewpoints on ethnography, landscape archaeology and cultural landscapes are very poorly developed in South Africa when compared to other parts of the world. The concept of geoheritage is thus also poorly developed and should be considered as an important future research theme linking together aspects of the physical and human environments (Scheermeyer 2005). Geoheritage issues are also at the heart of environmental and resource management, which is important in South Africa as a consequence of climate change and increasing pressure on soils, water and ecosystem services. The relationship between geoheritage and wider management issues, however, has not been well articulated.

19.3 Legislative and Management Context of Landscapes and Geoheritage Sites in South Africa

South Africa is regarded as having fairly progressive and innovative environmental legislation that recognises the natural, physical, economic and psychological importance of the environment to humans and vice versa. The National Environmental Management Act (NEMA), 107 of 1998 (amended 2013), is the key Act that sets out the principles for decision-making on matters affecting the environment; it also enables cooperative governance and provides for the administration and enforcement of other environmental management laws. These include the National Environmental Management: Protected Areas Act, 57 of 2003

Table 19.1 Sites declared as World and Provincial Heritage Sites in South Africa, with direct or indirect landscape/landform appeal and value

Site #	Site name and location	Site type	Heritage classification	Notes
1	**Horseshoe Waterfall**; Pilgrims Rest District, Mpumalanga Province	Natural	Provincial heritage site	Fluvial landform
2	**MacMac Waterfalls**; Pilgrims Rest District, Mpumalanga Province	Natural	Provincial heritage site	Fluvial landform
3	**Lone Creek Waterfall**; Pilgrims Rest District, Mpumalanga Province	Natural	Provincial heritage site	Fluvial landform
4	**Berlin Waterfall**; Pilgrims Rest District, Mpumalanga Province	Natural	Provincial heritage site	Fluvial landform
5	**Natural Rock Bridge**; Pilgrims Rest District, Mpumalanga Province	Natural	Provincial heritage site	Fluvial landform
6	**Echo Caves**; Lydenburg District, Mpumalanga Province	Natural	Provincial heritage site	Karst landform
7	**Dwars River Geological occurrence**; Lydenburg District, Mpumalanga Province	Natural	Provincial heritage site	Geology
8	**Goedehoop Natural Rock Bridge**; Ermelo District, Mpumalanga Province	Natural	Provincial heritage site	Fluvial landform
9	**Sterkfontein and other Caves**; Cradle of Humankind, Gauteng Province	Archaeological	World heritage site	Hominid fossils, artefacts, karst landforms
10	**Confidence Reef**; Roodepoort District, Gauteng Province	Natural	Provincial heritage site	Geology
11	**Vredefort Dome**; Parys District, Free State Province	Geological	World heritage site	Geology, landscape
12	**Glacial Pavement**; Durban, KwaZulu-Natal Province	Geological	Provincial heritage site	Geology, glacial landform
13	**Lynmouth Glacial Pavement**; Richmond District, KwaZulu-Natal Province	Geological	Provincial heritage site	Geology
14	**Ukhahlamba/Drakensberg Park**; KwaZulu-Natal and Lesotho border region	Natural/Cultural	World heritage site	Cultural landscape
15	**iSimanaliso Wetland Park**; St Lucia District, KwaZulu-Natal	Natural	World heritage site	Wetland, landscape
16	**Cretaceous Deposits**; Mzamba Beach, Mbizana District, KwaZulu-Natal Province	Geological	Provincial heritage site	Geology
17	**Howick Waterfall**; Howick District, KwaZulu-Natal Province	Natural	Provincial heritage site	Fluvial landform
18	**Cape Winelands**; Western Cape Province	Cultural landscape	Provincial heritage site	Cultural, landscape
19	**Geological Exposure**; Sea Point, Cape Town, Western Cape Province	Geological	Provincial heritage site	Geology
20	**Cango Caves**; Oudtshoorn District, Western Cape Province	Geological	Provincial heritage site	Karst landform
21	**Langebaanweg Quarry fossil site**; Langebaan, Western Cape Province	Palaeontology	Provincial heritage site	Fossils, palaeo-landscape
22	**Dumortierite occurrence**; Kenhardt District, Northern Cape Province	Geological	Provincial heritage site	Geology
23	**Nooitgedacht Glacial Pavement**; Kimberley District, Northern Cape Province	Geological	Provincial heritage site	Geology, glacial landform
24	**Driekops Glacial Pavament and Rock Engravings**; Herbert District, Northern Cape Province	Archaeological	Provincial heritage site	Archaeology, glacial landform
25	**Bucklands Glacial Pavement**; Herbert District, Northern Cape Province	Geological	Provincial heritage site	Geology, glacial landform
26	**Blaauboschdrift Glacial Pavament**; Herbert District, Northern Cape Province	Geological	Provincial heritage site	Geology, glacial landform
27	**Orbicule Koppie**; Namaqualand, Northern Cape Province	Geological	Provincial heritage site	Rare rock type, orbicule granite

Fig. 19.2 **a** A natural rock bridge in the Eastern Escarpment region of Mpumalanga, and **b** accompanying marker stone and information plaque, dating to 1977. The text on the uppermost plaque is in Afrikaans and English. The English version reads: 'This rock bridge is a rare geological phenomenon and originated through the dissolving of portions of the dolomite in ground water and the subsequent lowering of the surrounding land surface.' (*Photographs* S. Grab)

(amended 2009), National Environmental: Biodiversity Act, 10 of 2004, and the World Heritage Convention Act, 49 of 1999. South Africa's geoheritage is thus granted protection by virtue of being integral to a protected area, and/or through statute that restricts activities that could destroy, damage or deplete this resource.

WHS are protected through the World Heritage Convention Act no. 49, 1999. This Act provides for the incorporation of the World Heritage Convention into South African law and the establishment of a World Heritage Authority to safeguard the integrity of WHS. In terms of this legislation, the Authority must manage the site in accordance with all applicable national, provincial and municipal legislation, bylaws, policies and management plans.

The National Environmental Management: Protected Areas Act (NEMPA) as amended in 2009 identifies several types of protected areas. These are: (a) special nature reserves, national parks, nature reserves and wilderness areas; (b) WHS; (c) marine protected areas; (d) specially protected forest areas, forest nature reserves and forest wilderness areas; and (e) mountain catchment areas. In 2005, the Minister of the Department of Environmental Affairs and Tourism published the Regulations for Proper Administration of Special Nature Reserves, National Parks and WHS (GG 28181, GNR 1061, 28 October 2005). However, because these regulations were designed with state-owned protected areas in mind (Kotzé and de la Harpe 2008), and many of the conservation areas in South Africa are privately owned—for example large tracts of the FHSSA World Heritage Site and the Vredefort Dome—enforcement of regulations is very difficult. In most cases, the protection and management of state-owned land is undertaken by South African National Parks (SANParks) which administers 20 National Parks across the country, covering a range of different climatic and geologic/geomorphic settings and ecological environments; these encompass over 3.7 million ha in total. The legislative framework for National Parks is the National Parks Act, Act No. 56 (1926), and National Environmental Management: Protected Areas Amendment Act 15 (2009). In addition, there are a number of Transfrontier Parks in which larger protected areas cross international borders, such as |Ai-|Ais/Richtersveld Transfrontier Park (South Africa/Namibia), Kgalagadi Transfrontier Park (South Africa/Botswana) and Maloti–Drakensberg Transfrontier Conservation and Development Area (South Africa/Lesotho). Both National and Transfrontier parks have landscapes of significant geoheritage interest within them, including, for example, coastal sand dunes and the associated Duneveld bioregion, and hunter-gatherer archaeology of the Namaqua National Park on the west coast of South Africa.

Fig. 19.3 Photograph of geological garden at the Walter Sisulu National Botanical Garden, Johannesburg (*Photograph* S. Grab)

South Africa's geoheritage is also protected through various restrictions placed on activities that impact negatively on specific geo-landscapes or geo-objects, the latter including fossils, precious stones, minerals and meteorites. For example, the Nature Conservation Ordinance (No 12 of 1983) places restrictions on human activities in caves that include the removal of speleothems, stalactites and stalagmites. The Minerals and Petroleum Resources Development Act (MPRDA), Act 28 of 2002, also places restrictions on mining and quarrying for precious stones and mineral resources. The promulgation of the MPRDA removed the rights or privileges of landowners to determine who could have access to their land, and to what extent their land could be prospected or mined and granted the State the power to regulate every aspect of the exploitation of minerals (van der Schyff 2012).

Lastly, the National Heritage Resources Act, 25 of 1999, allows for the proclamation of archaeological and palaeontological sites and grants blanket protection to these sites, their materials and meteorites. Section 35 of this legislation makes these sites and objects the property of the state and fines or imprisonment may result from the destruction, damage, removal or trade in these materials. This Act also sets out the criteria and procedures for an Archaeological/Palaeontological/Heritage Impact Assessment that forms part of the Environmental Impact Assessment (EIA) process put into effect by NEMA. The EIA is intended to inform and guide decision-makers when the environment is to be impacted by mining or development. NEMA's broad definition of the environment makes provision for a set of integrated studies. Unfortunately, in practice, the interconnection between the natural and human environment is seldom made, because the studies are carried out by independent consultants and the reports are often reviewed by different departments at the national or provincial level, or even by different authorities. The SAHRA or their provincial counterparts review the heritage, archaeological and palaeontological reports, while the social and environmental aspects of the report are reviewed within the Department of Environment and their provincial counterparts.

19.4 Examples of Geoheritage and Geotourism in South Africa

19.4.1 West Coast Fossil Park

The West Coast Fossil Park at Langebaan, 120 km north of Cape Town (Fig. 19.1), is one of the world's richest Mio-Pliocene vertebrate faunal sites. The park (14 ha in extent) is one of South Africa's most important geoheritage/geotourist sites and was declared a National Monument in 1998 and National Heritage Site in 2005. A range of scientific studies has been undertaken on this site, including palaeontology, sedimentology, stratigraphy and geomorphology (e.g. Roberts et al. 2011). A museum, education centre and covered excavation site are open to the public and displays integrate components of palaeontology, archaeology, palaeoecology, sedimentology and landscape change. Models of regional hydrogeomorphic and climate changes are explained in terms of changes in sea level, palaeoecology, and fauna and flora. A particular management objective is to enhance and support regional education, research and tourism.

19.4.2 The Drakensberg San Rock Art Sites

The uKhahlamba–Drakensberg Park, which spans 300 km of the Great Escarpment, has the highest density of rock art sites in South Africa, and one of the highest globally, with over 35,000 individual rock art images recorded from over 600 sites (Lewis-Williams 1983). The art is a legacy of San hunter-gatherers who were exterminated by the mid- to late-nineteenth century in this region. The art depicts elements of the spiritual world drawn in trance, including images of wild fauna and even of the settler and hunter-gatherer conflicts. In recognition of this cultural heritage and to further develop heritage tourism in the park, the Kamberg and Didima Rock Art Centres were established in 2002 and 2003, respectively. In addition, the older Main Caves site at Giant's Castle Game Reserve includes a reconstructed San family group exhibition with San artifacts. Guided tours operate at these venues, but emphasis is primarily on the rock art. The cave sandstone overhangs where most of the rock art occurs; the weathering processes contributing to rock art decay and its associated preservation (Hoerle 2005), and the way in which the San used the diverse landscape (to trap wildlife, select family shelters, seek protection, etc.), are unfortunately not adequately incorporated into museum displays (Duval and Smith 2013). According to Mazel (2008, p. 41), these and other attributes of the natural and archaeological history should thus be considered as part of 'the interpretive requirements of the region as a whole'. Here lies a great opportunity to present the cultural- and closely connected geoheritage in an integrated manner.

19.4.3 The Makapan Valley Heritage Site

Makapan Valley, named after Mokopane, the chief of the Kekana people, was listed as part of the FHSSA World Heritage Site in 2005 and is located in Limpopo Province, north-east South Africa. The Makapan Valley is exceptional in that the caves within the valley preserve a continuous record of hominin occupation from 4 million years ago to the present (e.g. Esterhuysen 2010), and cave speleothems have provided detailed information about climate and environmental changes in the region. The richness of the Makapan fossil, archaeological and historical record is the result of its location at a point of intersection between distinct geologies and distinct ecologies. The Makapan–Strydpoort highlands area—the quartzite escarpment and deeply eroded dolomitic uplands—form part of the 'Wolkberg hotspot'. The rich and varied vegetation covering the highlands includes the tropical flora of a northern origin, and a more temperate flora of a southerly provenance (Fig. 19.4).

To date, a number of plans have been drafted to monitor and manage this heritage site and create a viable tourist destination. However, a range of issues has prevented this from becoming a safe and successful destination. The caves, in particular, are difficult to manage. For instance, dolomite

Fig. 19.4 Photographs of the Makapan Valley. **a** View down the Mwaridzi arm of the Makapan Valley showing the exposed Black Reef Quartzite (red cliffs) with lower lying dolomites. **b** View inside the Lime Works Cave where the roof supports have started to twist and lean due to movements within the cave system (*Photographs* A.B. Esterhuysen)

caves are prone to collapse. At most tourist sites of this nature, measures are taken to stabilise the caves or monitor them through the installation of 'early warning' systems. In the case of Makapan, blasting from a nearby diamond mine has added appreciably to the instability of the valley's caves. Furthermore, subsidence resulting from the extraction of water from the dolomite aquifers, for irrigation and domestic purposes, has added to the unpredictability of the system. Histoplasmosis, a potentially fatal disease usually associated with the inhalation of fungal spores from cave soils, is also difficult to mitigate against. Other impacts that hinge on effective and efficient management include fire management, the cultural use of resources and access to the valley. Here again, efforts have fallen short—on at least two occasions fire has destroyed all tourist infrastructure; the harvesting of wood and medicinal plants is increasing and unmonitored; and theft is an ongoing problem due to the insecurity of land tenure in the immediate area.

19.5 Summary and Outlook

Tourism is an important driver of economic development in rural South Africa, but the potential of geoheritage tourism has not yet been fully exploited (Reimold et al. 2006). A possible reason for this may be the uncertain and sometimes contentious relationship between the need to protect natural resources *vs* the drive for economic development. Geoconservation in rural areas, however, can also promote more effective ecosystem conservation, biodiversity and water management, and protection of other cultural heritage features including archaeology. Constraints imposed by legislation regarding the preservation of specific features of heritage or geo-object interest suggest that broader landscape-scale geoheritage management is more easily achievable and can link to other landscape attributes. With South Africa's declining mining sector and slow industrial growth, there is considerable pressure to further develop tourism for economic growth. This provides an opportunity for the development of geoheritage sites, or possibly even Geoparks, outside of the National Parks and well-known tourist routes, but these may be hindered by site-specific sociocultural and land management issues which may make them vulnerable to future development. Possible lesser-known sites that might be considered for geoheritage tourism or for Geopark inscription include the 'Valley of Thousand Hills' (inland of Durban), large parts of the Karoo, and the Vredefort and Tswaing meteorite impact sites.

Acknowledgments We thank reviewers Murray Gray and Lars Erikstad for their comments on an earlier draft of this chapter.

References

Bailey GN, Reynolds SC, King GCP (2011) Landscapes of human evolution: models and methods of tectonic geomorphology and the reconstruction of hominin landscapes. J Hum Evol 60:257–280

Brocx M, Semeniuk V (2007) Geoheritage and geoconservation—history, definition, scope and scale. J R Soc W Aust 90:53–87

Cairncross B (2011) The National Heritage Resource Act (1999): can legislation protect South Africa's rare geoheritage resources? Resour Policy 36:204–213

Chirikure S (2013) Heritage conservation in Africa: the good, the bad, and the challenges. S Afr J Sci 109(1/2), Art. #a003, p 3. doi:10.1590/sajs.2013/a003

Cotterill FPD, de Wit MJ (2011) Geoecodynamics and the Kalahari epeirogeny: linking its genomic record, tree of life and palimpsest into a unified narrative of landscape evolution. S Afr J Geol 114:489–514

Dowling RK (2011) Geotourism's global growth. Geoheritage 3:1–13

Duval M, Smith B (2013) Rock art tourism in the uKhahlamba/Drakensberg World Heritage Site: obstacles to the development of sustainable tourism. J Sust Tourism 21:134–153

Erikstad L (2013) Geoheritage and geodiversity management—the questions for tomorrow. Proc Geol Assoc 124:713–719

Esterhuysen AB (2010) Excavation at Historic Cave, Makapan's Valley, Limpopo: 2001–2005. S Afr Arch Bull 65:67–83

Gray M (2013) Geodiversity—valuing and conserving abiotic nature. Wiley, Chichester, 495 pp

Gray M, Gordon JE, Brown EJ (2013) Geodiversity and the ecosystem approach: the contribution of geoscience in delivering integrated environmental management. Proc Geol Assoc 124:659–673

Hoerle S (2005) A preliminary study of the weathering activity at the rock art site of Game Pass Shelter (KwaZulu-Natal, South Africa) in relation to its conservation. S Afr J Geol 108:297–308

Kotzé L, de la Harpe S (2008) The Good, the bad and the ugly: using good and cooperative governance to improve environmental governance of South African World Heritage sites: a case study of the Vredefort Dome. Potchefstroom Electron Law J Potchefstroomse Elektroniese Regsblad 2:1/53, doi:10.4314/pelj.v11i2.42233

Lewis-Williams JD (1983) The rock art of southern Africa. Cambridge University Press, Cambridge, 68 pp

Mazel A (2008) Presenting the San Hunter-Gatherer past to the public: a view from the uKhahlamba-Drakensberg Park, South Africa. Conserv Manag Archaeol Sites 10:41–51

Pereira P, Pereira D (2010) Methodological guidelines for geomorphosite assessment. Indications méthodologiques pour l'évaluation des géomorphosites. Géomorphologie: relief, processus, environnement 2010:215–222

Reimold WU (1999) Geoconservation—a southern African and African perspective. J Afr Earth Sci 29:469–483

Reimold WU, Whitfield G, Wallmach Th (2006) Geotourism potential of southern Africa. In: Dowling R, Newsome D (eds) Geotourism. Butterworth-Heinemann, Oxford, pp 42–62

Reynolds SC, Bailey GN, King GCP (2011) Landscapes and their relation to hominin habitats: case studies from Australopithecus sites in eastern and southern Africa. J Hum Evol 60:281–298

Roberts DL, Matthews T, Herries AIR, Boulter C, Scott L, Dondo C, Mtembi P, Browning C, Smith RMH, Haarhoff P, Bateman MD (2011) Regional and global context of the Late Cenozoic Langebaanweg (LBW) palaeontological site: West Coast of South Africa. Earth-Sci Rev 106:191–214

Ruban DA (2010) Quantification of geodiversity and its loss. Proc Geol Assoc 121:326–333

Ruddiman WF (2012) Bridging a disciplinary gap. In: Giosan L, Fuller DQ, Nicholl K, Flad RK, Clift PD (eds) Climates, landscapes, and civilizations. AGU geophysical monograph series, vol 198, pp 1–10
Scheermeyer C (2005) A changing and challenging landscape: heritage resources management in South Africa. S Afr Arch Bull 60:121–123
Schutte IC, Booysen I (2010) Geosite identification as an element of promoting geotourism in a wildlife hotspot. In: Dowling RK, Newsom D (eds) Global geotourism perspectives. Goodfellow Publishers, Oxford, pp 124–136
van der Schyff E (2012) South African mineral law: a historical overview of the State's regulatory power regarding the exploitation of minerals. New Contree 64:131–153
Zunckel K (2003) Managing and conserving southern African grasslands with high endemism—the Maloti-Drakensberg Transfrontier Conservation Development Program. Mt Res Dev 23:113–118

Index

S. Grab and J. Knight (eds.), *Landscapes and Landforms of South Africa*,
World Geomorphological Landscapes, DOI 10.1007/978-3-319-03560-4,

C

Q

T

MIX
Papier aus verantwortungsvollen Quellen
Paper from responsible sources
FSC® C105338

If you have any concerns about our products,
you can contact us on
ProductSafety@springernature.com

In case Publisher is established outside the EU,
the EU authorized representative is:
Springer Nature Customer Service Center GmbH
Europaplatz 3, 69115 Heidelberg, Germany

Printed by Libri Plureos GmbH
in Hamburg, Germany